200 ILLUSTRATED SCIENCE EXPERIMENTS FOR CHILDREN

ROBERT J. BROWN

TAB Books
Division of McGraw-Hill, Inc.
New York San Francisco Washington, D.C. Auckland Bogotá
Caracas Lisbon London Madrid Mexico City Milan
Montreal New Delhi San Juan Singapore
Sydney Tokyo Toronto

To Bob Brown, Jr., Mary Brown, Betty Brown Rushing, and Barbara Black.

FIRST EDITION
FIFTEENTH PRINTING

Library of Congress Cataloging-in-Publication Data

Brown, Bob, 1907-
200 Illustrated science experiments for children.

Includes index.
1. Science—Experiments. I. Title. II. Title:
Two hundred illustrated science experiments for children.
Q164.B8415 1987 507′.8 86-23196
ISBN 0-8306-2825-8 (pbk.)

NS
2825

Line drawings by Frank W. Bolle. Technical assistance by Arthur Wood.

Contents

NOTICE

For children's use only with adult supervision.

Use chemicals or electricity only under adult supervision.

Use fire only with adult supervision.

Keep chemicals off of skin. Wash chemical apparatus.

Follow directions. Be careful. A chemical laboratory at home can be dangerous, but it can be lots of fun.

Chemicals can be poisonous even when absorbed through the skin.

Any cut in the skin should be washed immediately with plenty of water.

Don't work alone. Have someone with you, ready to turn off a switch if the experiment goes wrong—and this can happen with the best of us.

Introduction

Many people will enjoy and profit by this book. There are the young, eager to learn and to try new experiences. There are the teachers who are looking for simple yet scientifically sound experiments for their classrooms. There are the scientists who like to venture into the fun side of their occupations, and there are the members of the general public who like tricks that amuse and perhaps add a little to their stock of scientific knowledge.

I say tricks. One chapter in the book is headed "Tricks"—but actually almost the entire book is made of tricky stunts and experiments. In such a mundane observation as water gurgling from a jug, a tricky explanation is necessary. The "Bottle Imp" experiment is a trick, yet it is part of the "Air, Sound, Vibrations" chapter.

Some of the tricks are suitable for the very young. A simple electromagnet can be made by a six-year-old. Yet some are of such nature that they may be used as science fair projects by pupils in high school. An example is "Light Under Water," which can be constructed easily. Yet if it is to be made to go up or down, as suggested, it becomes a very difficult job, one worthy of a science fair first prize if made to work.

Some experiments have been explained erroneously elsewhere, and these have been included here with their correct explanations. Examples of this category are "Reading the Meter" and "Open Circuit, Short Circuit," the latter being terms that are often confused.

There are some experiments that, to my knowledge, cannot be explained adequately even on the lower college level. The "Mystery Color Wheel" is an example. Yet it can be made easily by elementary grade pupils, who will find it most fascinating.

Many common occurrences in the home can be considered good science experiments, and some of these are used in this book. Examples are "Butter Making," "Mother of Vinegar," and a "Coconut Culture." Steam, vapor, and gas are common to everyone, and while merely observed, they can be explained and considered an experiment.

I have striven for accuracy. Each experiment in this book was first set up by a boy or girl and is considered *only if it works*. Many experiments do not. Photographs are made, drawings are made from the photographs, then three very thorough consultants, one a professor of physics, another a professor of chemistry, and the third a scientist at the Oak Ridge National Laboratory, check for accuracy in the presentation and explanation.

Only then does the experiment go out to the syndicate for use in my column "Science for You" which is featured in newspapers throughout the United States and Canada. Errors are few; usually they are local typographical mistakes.

So, here's my invitation to get our pots and pans, balloons and candles, salt and vinegar, and start having fun with science. While having fun, it is just possible that some fundamental knowledge may be gained, too.

1

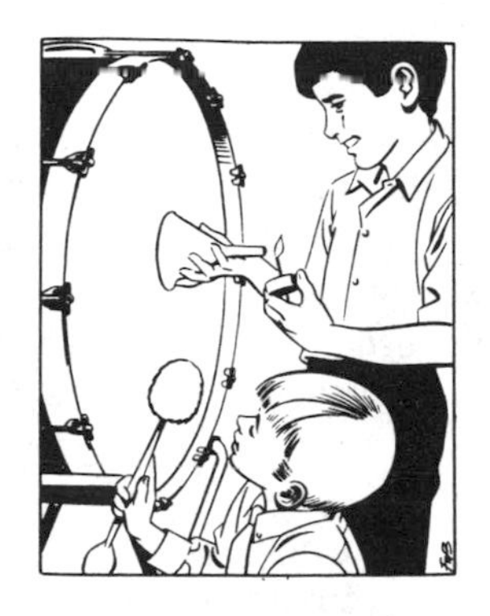

Air, Sound, Vibrations

THE BOUNCING BALLOON

Needed: Balloons and smaller rubber balls.

Do This: Put a ball into a balloon, blow air into the balloon, and hold the balloon neck end down. The ball will fall into the neck, close it, and hold the air. Let the balloon fall to the floor, and it will bounce several times, often quite high.

Here's Why: For every action there is an equal and opposite reaction (Newton's law). A force is acting inside the balloon to make the air rush out of the hole as the ball bounces up, and so there must be a force acting in the opposite direction. The last force causes the balloon to move upward as the air rushes out downward.

Another Explanation: The balloon with its end closed off by the ball contains air pressure equal in all directions. All forces due to internal pressure are balanced, up and down, right and left, etc.

As the ball hits the floor and bounces away from the hole in the balloon, some air rushes out. It cannot push against a hole, but on the opposite side it pushes against the rubber all over.

Thus, the air pushes on more area on the side opposite the hole, and since the push is the same per unit area, the force, which is pressure times area, is greater on the side opposite the hole, and the balloon moves away from the hole.

COLD AIR AND WARM AIR

Needed: A balloon and a refrigerator.

Do This: Blow up the balloon, tie it tightly so that there is no leak, and measure around it with a tape measure. Keep the balloon in the refrigerator half an hour, then measure it again. It will have shrunk in size.

Here's Why: As the air in the balloon is cooled, it contracts, exerting less pressure on the tight rubber of the balloon; consequently the balloon shrinks. When I tried it, the balloon shrunk from 24 inches to 21 inches in circumference.

TO PROVE AIR HAS WEIGHT

Needed: Two balloons, a yardstick.

Do This: Inflate two balloons and tie them to the ends of a yardstick. Balance the stick by moving the paper clip along it (clip is shown in the drawing). Puncture one of the balloons and the stick will no longer balance.

NOTE: This experiment is often presented as proof that air has weight. It does not provide this; it does prove that compressed air has greater weight than normal air.

This experiment has been presented using plastic bags instead of rubber balloons. It will not work, simply because little or no pressure can be put into a plastic bag. The plastic bags will continue to balance whether punctured or not, because the air inside and outside of them will continue to have the same pressure.

AIR EXPANSION

Needed: A fruit jar with tightly fitting lid, balloon, pan of water, a stove.

Do This: Place a few tablespoons of water in the jar, place the jar in the pan of water, and bring it to a boil. Put some air into the balloon, tie it tightly, drop it into the jar, tighten the lid on the jar, and let the jar cool rather quickly. The balloon will expand and perhaps fill the jar.

Here's Why: The steam had driven most of the air from the jar, and when it cooled and changed back into water by condensation, it occupied very little space. A gas (and the same applies to air, a mixture of gases) will fill any space in which it is confined.

To make this work better and faster, dissolve as much salt as possible in the water in the pan. Then when it boils it will be hotter than the boiling point of the plain water in the jar, and this will make the water in the jar boil more quickly.

WHY CAN WE SEE OUR BREATH ON A COLD DAY?

Here's Why: When air comes up from the large moist surfaces in the lungs, it is warm and holds much moisture. Since cold air cannot hold as much moisture as warm air, some of the moisture in the breath condenses into tiny droplets as the breath becomes cooler.

AN AIRFOIL

Needed: Cardboard, cellophane tape, a vacuum cleaner hose.
Do This: Tape the card to a larger card as shown, so the shape

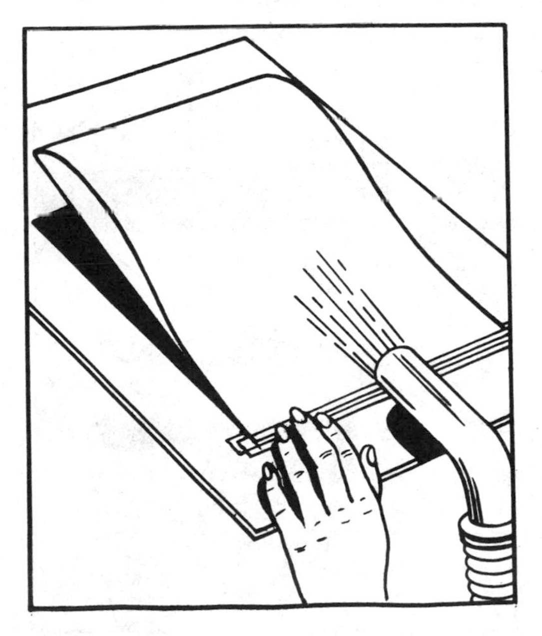

of the upper card is that of an airplane wing or "airfoil." When air is blown over it, the lift causes it to rise.

Here's Why: Bernoulli discovered that when air moves faster it has lower pressure. The air flowing over the card flows faster than the air below, which in this case is almost still. In the airplane wing or "airfoil" the upper surface is curved, making it larger than the flatter lower surface. The lift is on the upper surface.

THE COANDA EFFECT

Needed: A lighted candle, soda straw, jar or can.

Do This: Lie on the floor, hold the straw at the side of the glass (as shown in the drawing), and blow. The air stream curves around the outside of the jar and can blow out the candle after making as much as a quarter turn.

Here's Why: Henri Coanda discovered this effect when he built and flew a flying machine in 1910. It is the tendency of a fluid to follow the wall contour when discharged adjacent to the surface. The increased velocity of the air near the jar causes a decrease in pressure, so that the pressure of the still air keeps the moving stream near the glass until the velocity decreases.

The effect is noticed when water is poured from a glass; it tends to run down the side. Here the water is attracted to the surface also by adhesion or surface tension.

Do This: Place the ball between the boards, blow air into it through the hose, and it will lift a heavy person.

Here's Why: If air at only 1 pound per square inch is blown into the ball, it will lift 100 pounds if 100 square inches of ball touch the upper board. This is one of the laws of hydraulics, and explains how 200 pounds of air from a compressor in a filling station can lift a 3,000-pound automobile (Pascal's law).

SOUND CONDUCTION

Needed: A yardstick.

Do This: Hold the stick so that one end is against the ear. Reach out as far as possible and scratch the stick. The sound will be heard distinctly from the stick. Now hold the stick away from the ear. Reach out the same distance as before and scratch it. The sound will be much weaker.

Here's Why: Sounds are waves traveling in an elastic medium, usually air. But solids are better conductors of sound; therefore there is better conduction of sound through the stick to the ear than through the air between the stick and the ear as in the second action.

Yes, the stick is elastic.

So is steel; so is glass; so is water—in varying degrees.

THE "BOTTLE IMP"

Needed: A bottle, small piece of paper, soda straw.

Do This: Place the bottle horizontally on a table. Roll the pa-

per into a small ball, wetting it slightly to make it keep its shape. Place the ball in the neck of the bottle. Blow into the bottle and the "imp" will force the ball out.

Here's Why: When the breath is blown toward the bottle, some air is likely to go past the ball because the air behind the ball, plus its inertia, tend to keep it in place. The compressed air behind the ball then forces it out.

Now, blow through the straw, directly *on* the paper ball, and it can be blown into the bottle. Here the force of the blown air is directed on the paper and forces the ball into the bottle before additional air gets in behind it.

TIMBRE

Needed: Two people.

Do This: Have one person hum while the other whistles the same note, or have one hum while the other plays the same note on an instrument. They cannot sound the same.

Here's Why: A pure musical sound would comprise only a single vibration at a definite frequency, and only some electronic instruments make such sounds. Most other tones, such as those made by lips or throat or a musical instrument, contain not only the main

or "fundamental" vibration but many varied overtones—vibrations at higher frequencies than the fundamental.

The mixture of these vibrations gives quality or "timbre" to the sound. This is why we can tell whether it is Mother's voice or Aunt's. This is why we can tell whether the instrument we hear is a violin or a harp.

THE SOUNDING BOARD

Needed: A comb.

Do This: Rub your fingernail along the teeth of the comb and listen to the sound. Hold the end of the comb against a door panel or table top, rub the teeth the same way and note the louder sound.

Here's Why: As the teeth are rubbed, they vibrate and transmit their vibrations to the air around them. When the comb is held against the door, it transmits its vibrations to the door as well as to the air. The vibrating wood transmits vibrations to a larger volume of air and may be heard as a louder sound.

This is why a piano and many other instruments have sounding boards. Their sound would not be as loud without the additional vibrations of the "boards."

MELLOW THE RADIO

Needed: A pocket-size radio, jars, tubes. (I had best results using a wide-mouth quart fruit jar, an oatmeal box, and a tube that held Christmas wrapping paper.)

Do This: Try holding the radio, speaker side down, over various size jars and tubes. The sound will be different with each one. Some jars will amplify the bass tones, making the sound much more mellow.

Here's Why: A resonance is set up when the radio is placed on a jar or tube. Some low frequency sound waves join with their reflected waves so that they reinforce each other, making the low-frequency sounds louder.

HEAR THE DOPPLER EFFECT

Needed: An alarm clock, strong string.

Do This: Tie the clock securely to the end of the string. Start the alarm. Then whirl the clock around. To someone listening nearby the pitch of the alarm will be higher as the clock approaches on its swing around, and lower as it moves away from the listener.

Here's Why: Frequency or pitch is determined by the number of sound waves striking the ear per second. The clock alarm gives a constant number, but as it moves toward the listener more sound waves reach the ear per second than when the clock moves away.

NOTE: This principle, discovered by Christian Johann Doppler in 1842, is used in astronomy to determine the rate of rotation of the sun. In radar it makes possible several speed-check devices used on the highways.

VIBRATIONS

Needed: A big drum, a big funnel, and a lighted candle.

Do This: Hold the funnel so that it concentrates the vibrations from the drum on the candle flame. When the drum is beaten, the vibrations can extinguish the flame.

Here's Why: The sound of the drum—and all sounds coming through air—is made up of back and forth motion of air. A high pitch is made of rapid but small movements. A low or deep tone of the same loudness is a slow but large back and forth movement of the air. The tone of a large drum is made of air movements so strong that when concentrated by the funnel they can blow out the candle flame. See top illustration, page 14.

CHLADNI FIGURES

Needed: A metal plate mounted by a bolt in the middle, a yardstick, cloth shoelace, sand, rosin. See bottom illustration, page 14.

ATMOSPHERIC PRESSURE

Needed: A pint jar with tight lid, candle, soda straw, water.

Do This: Make a hole in the lid, insert the straw, and drop wax around it to seal it. Screw the cap on the jar of water and try to suck water out through the straw. The water may be sucked out only when the lid has been loosened. Try blowing into the jar through the straw. When the mouth is taken away water will spurt up.

Here's Why: Due to its large pressure (more than 2,000 pounds per square foot) our atmosphere affects everything within it. When we try to suck water from the airtight jar, we reduce the pressure within and very little water will come out. The straw may collapse if we continue to suck on it due to the greater pressure on the outside. We can blow quite a bit of air into the jar since air is very compessible. The increased pressure of the air above the water in the jar will force the water up and out of the straw.

LIFT A HEAVYWEIGHT

Needed: A beach ball, flexible hose and connection, two planks hinged together at one end.

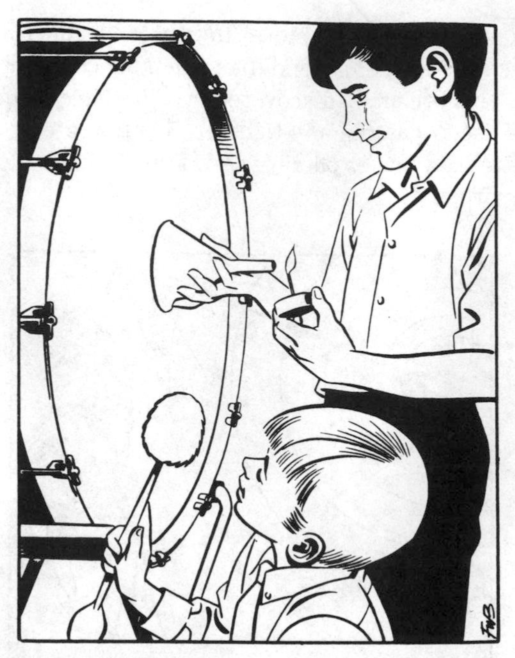

Do This: Mount the plate so that the edges are free to vibrate. Make a bow like a violin bow from the stick and shoelace, and rosin the lace thoroughly. Sprinkle sand on the plate, draw the string

quickly along its edge, and the sand grains will dance, settling into beautiful patterns.

Here's Why: Ernst F. Chladni, German physicist, discovered this way of showing vibration patterns. The sand settles along the nodes of the vibrating metal. The patterns are different when the bow is drawn along different parts of the plate.

THE GRASS WHISTLE

Needed: Blades of grass.

Do This: Hold the grass blade between the thumbs as shown, and blow through the opening. After a little practice a musical tone can be produced, and altering the shape of the hollow in the hands alters the pitch.

Here's Why: Air blowing across the blade of grass catches it on one side, blowing it slightly so that the air next catches it on the other side, blowing it back again. This back and forth motion, or vibration, produces the sound waves that are heard.

The edge of the blade of grass may move back and forth several hundred times each second.

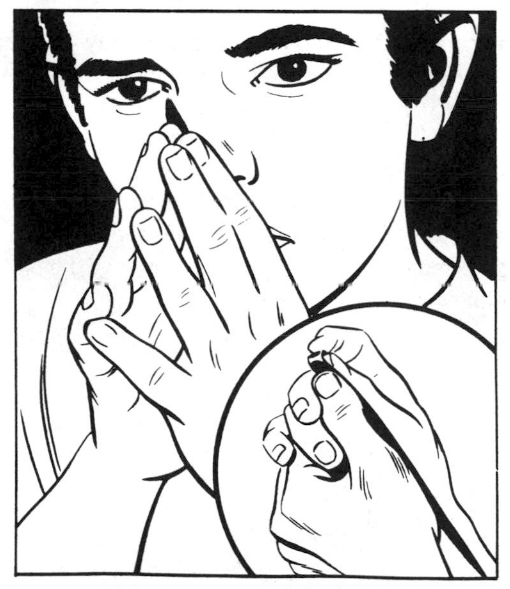

NODES AND ANTINODES

Needed: A rigid electric bell with gong removed, string.

Do This: Attach the string to the clapper of the bell as shown. As the bell "rings" hold the string in the hand below it, varying

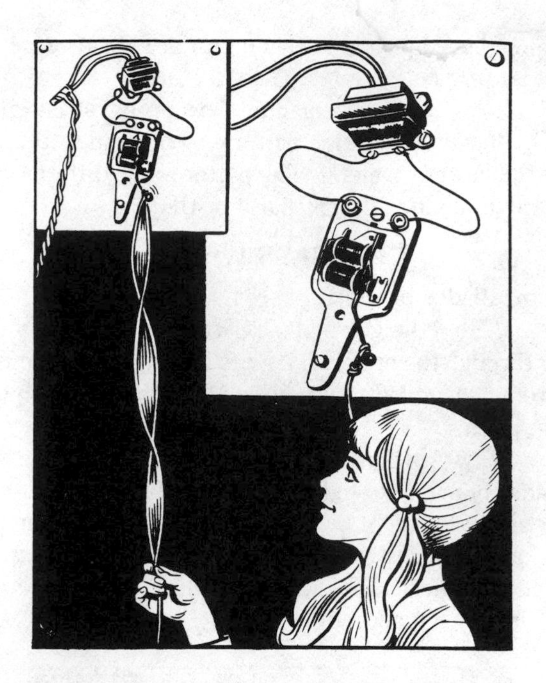

the tension and length. Vibration patterns may be created, with portions of the string (the antinodes or loops) vibrating widely and points (the nodes) remaining still.

Here's Why: Waves are traveling down the string from the vibrating bell clapper and up the string as they are reflected from the hand. At the nodes the initial and reflected waves cancel the effects of each other, while at the antinodes the two waves add their effects to give maximum motion to the string. These waves are called standing waves.

2

Water, Surface Tension

MOLECULAR MOTION

Needed: Two water glasses, water, salt, food color.

Do This: Put a pinch of salt into a glass of water, stir, then let the glass be still until the motion of the water has stopped. Put a drop of food color into a half glass of water, mix, then pour care-

fully into the glass of salt water. The colored water will remain mostly above the salt water because the salt water is heavier. Let the glass set undisturbed overnight. The liquids may have mixed.

Here's Why: The mixing occurs because the water molecules are in constant motion, bumping into each other, moving in all directions. This is molecular motion or Brownian motion, and it takes place in all liquids and gases. A familiar example is when an odor fills the air in a room even when there is no apparent circulation of the air.

NOTE: To pour the colored liquid into the salt water so it does not immediately mix, wet a piece of paper and place it on the salt water surface. Then pour the colored water slowly onto it.

If not enough salt is in the water, the mixing will be quick. If there is too much salt, the mixing may not take place even after several days of leaving the liquids undisturbed.

THE MYSTERIOUS PING-PONG BALL

Needed: A ping-pong ball with a string attached and a stream of water from the faucet.

Do This: Turn on the water, hold the ball by the string, and let the ball touch the moving stream of water. The ball will cling

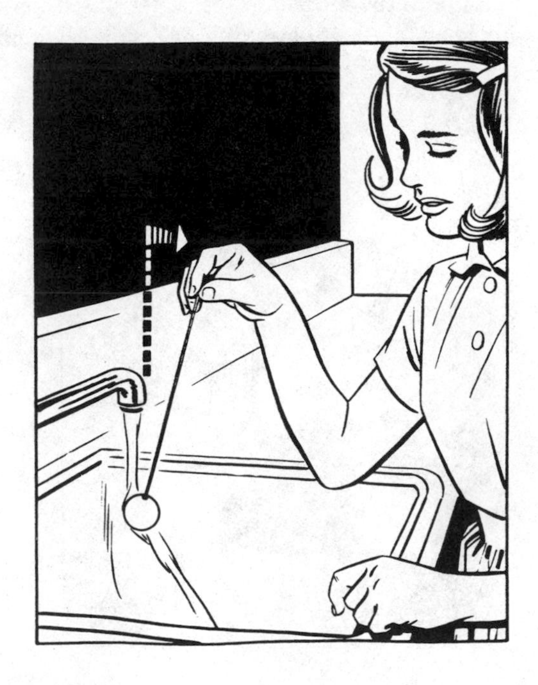

to the stream even if the string is moved outward to a considerable angle.

Here's Why: This is a variation of the Bernoulli principle in which the lateral pressure of moving air is less than that of the still air around it. Here the pressure of the water, which is moving around one side of the ball, is less than that of the still air on the other side, and as the ball attempts to move away from the moving stream of water, the air will push it back.

Adhesion is an important factor here, too. Because of adhesion the water tends to cling to the ball as it flows around it.

BERNOULLI EFFECT ON A WATER SURFACE

Needed: A soda straw, two toothpicks, a dish of water.

Do This: Place the toothpicks on the water, side by side, and blow gently through the straw so the air flows between them. The toothpicks move together.

Here's Why: According to the Bernoulli effect, pressure of a fluid, either water or air, is reduced as its speed is increased. The air pressure between the straws is reduced because it moves, and the water surface also is made to move by the moving air, and so its pressure, too, is reduced.

This experiment is not easy to do the first time because the force of the air acting to push the toothpicks apart may exceed the force of the Bernoulli effect.

FLOAT AN EGG

Needed: Two wide mouth jars, water, salt, an egg.

Do This: Fill one jar half full of water and stir in salt until no more will dissolve. Drop the egg into the salt water and it will float. Now pour plain water carefully over the egg and the jar may be filled while the egg still floats on the salt water, under the plain water.

Here's Why: According to Archimedes' principle, a body wholly or partially immersed in a fluid is buoyed upward with a force equal to the weight of the volume of liquid is displaces. Salt water is heavier than the egg and so the egg floats on it. But the egg is heavier than plain water and so does not float on it.

If the experiment is undisturbed, the egg will remain halfway down in the liquid for several days.

VISCOSITY

Needed: A glass of water, a glass of syrup, two marbles.

Do This: Drop one marble into each glass. Note that it moves downward much more slowly in the syrup.

Here's Why: Viscosity is defined as internal friction in fluids

due to adherence of particles to one another, or the resistance of a substance to being fluid because of molecular attraction. The particles adhere to one another in both water and syrup, but more so in the syrup. Therefore the marble has a more difficult time pushing through the syrup.

SHOW MOLECULAR MOTION

Needed: Two glasses of water, one hot and one cold, two soda straws, ink or dark food color.

Do This: Place a drop of ink in each straw. Let the water stand until it is still. Place the ends of the straws below the surface of the water, one in one glass and one in the other. Squeeze the straws until the ink comes out. The ink mixes more quickly with the hot water than it does with the cold water.

Here's Why: The molecules of the water are in violent motion, faster in the hot water than in the cold. The molecules in the ink are also in motion. The collisions of the molecules mix the ink with the water, and the mixing process is faster in the hot water where the molecular motion and collisions are faster.

Molecular motion of air is demonstrated when an odor spreads throughout a room.

A SALT GARDEN

Needed: Salt, water, a pan, pieces of brick or coal.

Do This: Mix salt and water in a jar, all the salt the water will

take. Then pour the clear solution into the pan. Place pieces of coal, brick, or earthenware in the solution so they stand up. In a day salt will be seen "growing" on the surfaces of the objects.

Here's Why: By capillary action the salt water is drawn up through small openings in the coal and brick until it reaches the top or sides. There the water evaporates into the air, leaving the salt behind.

NOTE: Charcoal briquettes have been recommended for this, but they are not usually satisfactory. A little ammonia in the water may make the water less oily and aid the capillary action. And a little ink or food coloring dropped on the accumulating salt crystals will give them color.

A CAPILLARY FILTER

Needed: A piece of discarded cotton sheet, jar of muddy water, a clean glass.

Do This: Wind the cloth into a roll, and place it in the muddy water so that one end is in the water jar and the other end hangs down into the empty glass. The empty glass must be considerably lower than the one with the muddy water. The water will slowly pass through the cloth into the lower glass and will be almost clear.

Here's Why: Capillary action occurs when a liquid "wets" another material. In this case the water wets the cloth due to the attraction between the unlike molecules. This attraction may be called adhesion.

The smaller water molecules pass along the small openings in and between the fibers of cloth, leaving behind the larger dirt and mud particles.

OSMOSIS

Needed: Salt, water, raisins or prunes, two glasses.

Do This: Put water into two glasses. In one glass only, add as much salt as will dissolve. Drop the dried fruit into both glasses. The fruit in the plain water will swell, while the fruit in the salt water will remain shriveled.

Here's Why: According to the laws of osmosis, a liquid will go through cell walls, mostly moving from the lesser concentrated solution to the one in which more substances are dissolved. The fruit, of course, is made up of cells, and normal tap water flows through their walls into the cells where the liquid is more dense. Since the salt water solution is heavier than the cells' liquid, little or no salt water will flow into the fruit.

WETTING THE WATER

Needed: Two glasses of water, string, scissors, detergent.

Do This: Add detergent to only one glass of water. Clip the string into pieces, dropping them on the water. The string will float on top of the plain water but will sink into the water and detergent mixture.

Here's Why: The detergent reduces the surface tension of the water so that it is less likely to support the weight of the string pieces. It also dissolves some of the oil and other materials on the string so that the string gets wet more easily. Detergent makes the water wetter.

If the string contains too much sizing or other substances, it may be necessary to wet it and wring out as much water as possible before trying the experiment.

OIL CHASER

Needed: Oil, water, a drinking glass, soap, toothpick.

Do This: Let a glass of water set until the surface of the water is still. Place two or three drops of oil on the water. The oil will cling together, forming a round spot.

Stick a toothpick into a small piece of soap and touch it to the middle of the oil spot. The oil is pulled toward the edge of the glass.

Wash the glass thoroughly. Again drop oil on the surface of the water. The oil may be moved around with a clean toothpick.

Here's Why: The surface tension of the water is balanced in all directions before the oil is added. Oil and water do not mix. When soap is added it dissolves somewhat in the water, lowering the surface tension. Therefore, the greater pull of the surface tension around the oil pulls the oil toward the edge of the glass.

SURFACE TENSION MEASUREMENT

Needed: Screen wire, a glass of water.

Do This: Cut the screen wire to form a box, as shown, and attach single wires from the corners of the box to form a handle. If the box is lowered into the water so it is covered, a film forms across the box end. Then as the box is pulled upward, it brings water with it for a considerable distance before the film breaks.

COMMENT: Surface tension can be measured this way. Cohesion and adhesion form the thin film between the wires—a film which is surprisingly strong.

Cohesion: force holding a solid or liquid together because of

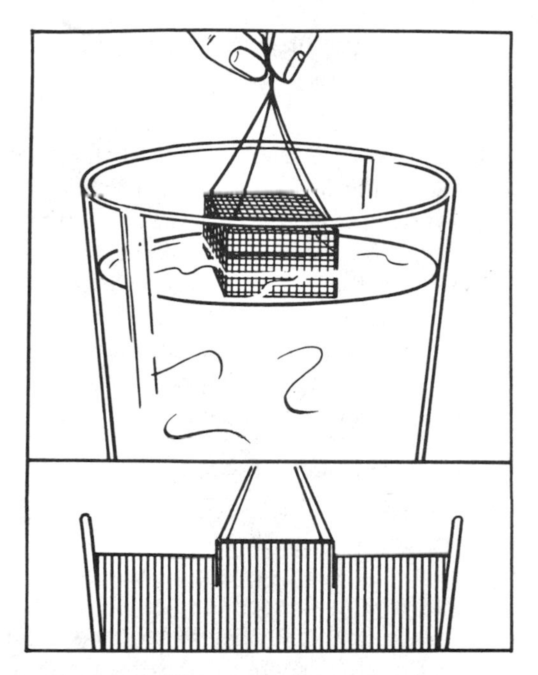

the attraction between like or similar molecules.

Adhesion: the effect of sticking to a surface, as produced by forces between unlike or dissimilar molecules.

A WATER "SMOKESTACK"

Needed: A jar with a metal screw top, ink or food color, water in a large tank or bucket, hot water in the jar.

Do This: Punch a hole in the metal top of the jar with a nail. Add coloring to the water in the jar. Hold a finger over the hole and place the jar into the tank of water. Take away the finger, and a stream of colored water rises like smoke from a smokestack.

Here's Why: The hot water is lighter in weight than the cold water in the tank, so it tends to rise as cold water takes its place in the jar. Try it with two holes in the jar top. The "smoke" should rise faster because a convection current will be set up in the liquid in the jar. See bottom illustration, page 27.

SURFACE TENSION (1)

Needed: A spoon and a water spigot.

Do This: Hold the spoon under the water flow, and the water can be made to form a sheet as it leaves the spoon.

Here's Why: Surface tension on the surface of water acts much as a stretched rubber sheet. It is able to hold the water together for some distance from the spoon. As the sheet is broken up the surface tension continues to act; it tends to keep the water in round drops, since a sphere is the form having the smallest surface for a given volume.

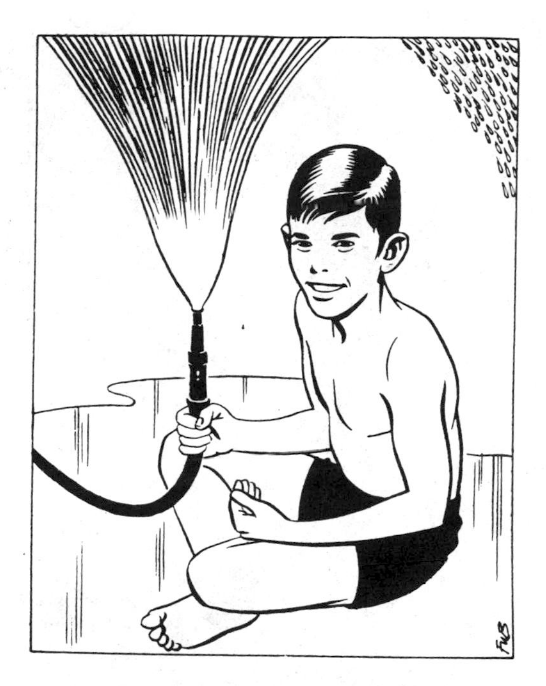

SURFACE TENSION (2)

Needed: A hose with adjustable nozzle, water.

Do This: Adjust the nozzle so a cone-shaped sheet of water is produced.

Here's Why: The surface of water acts much as a stretched rubber sheet. This surface tension holds the water together in the cone-shaped sheet until the diverging movement of the water pulls it apart.

Even then the surface tension remains in force, pulling the blobs of water into almost spherical drops. A sphere has the smallest surface area of any shape; friction with air as the drops fall distorts the shape somewhat.

DRY?

Needed: Dried beans or peas, a tall tin can, a piece of glass or a dish, a stove.

Do This: Place the beans or peas in the can and cover with the glass or dish. Put can on the stove, and heat it with mild heat. Notice that droplets of water will appear on the underside of the glass or dish.

Here's Why: Most substances we think of as being dry actu-

ally contain water, as do the beans or peas. The heat evaporates some of this water, and the water then condenses on the cool plate or glass. The experiment may work better if the beans are split or crushed slightly before being heated.

"SMOKE RINGS" IN WATER

Needed: A dish or jar of water, a drop of ink or colored water.

Do This: Drop the ink into the water, and in many cases it will form into rings resembling smoke rings.

Here's Why: As the drop hits the surface of the water its momentum carries it under the surface, setting up a rolling doughnut-shaped motion. In this motion the colored liquid goes downward, clear water coming in above it to take its place. The clear water makes the hole. But this motion cannot last long. The ink soon mixes with the water. See bottom illustration, page 30.

THE FLOW FROM A JUG (1)

Needed: A gallon jug, clock or watch with a sweep second hand, water.

Do This: Fill the jug half full with water. Put a hand over the neck, invert the jug over the sink, remove the hand quickly and see how long the water takes to gurgle out.

Fill again to the same point (mark the point with a grease pencil or tape) and this time start the water swirling before the hand is removed. The water will take longer to get out.

Here's Why: When the water swirls, its spin "feeds on" the

energy of the water flow and spins faster as the water pours out, until friction causes a balance. The water tries to keep its spin velocity as it reaches the narrow part of the jug, so it spins faster there, but its increasing centrifugal force tends to force it back into the larger diameters of the jug. If the spin could be fast enough, all the water would be forced into the big part of the jug and none would pour out.

THE FLOW FROM A JUG (2)

Needed: A gallon jug, clock or watch with a sweep-second hand, soda straw, water.

Do This: Fill the jug half full with water, invert it, and see how long it takes the water to gurgle out. Now do the same, but insert the soda straw so that its end extends above the surface of the water in the inverted jug. The water flows out more smoothly and more quickly.

Here's Why: The resistance to the flow of air through the straw is less than the gurgling flow of air because the air does not have to push through the water. The flow of water is faster than in the gurgling jug because once it builds up its downward motion (kinetic energy) it keeps it rather than stopping and backing up to let an air bubble through. The straw, even though small, can supply enough air to keep the pressure in the jug close to atmospheric pressure.

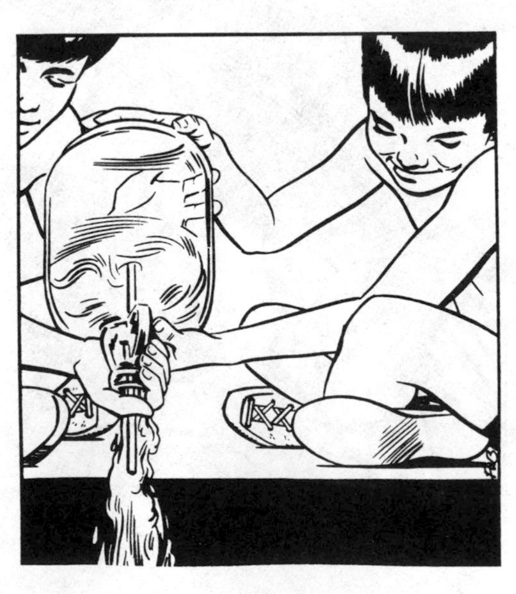

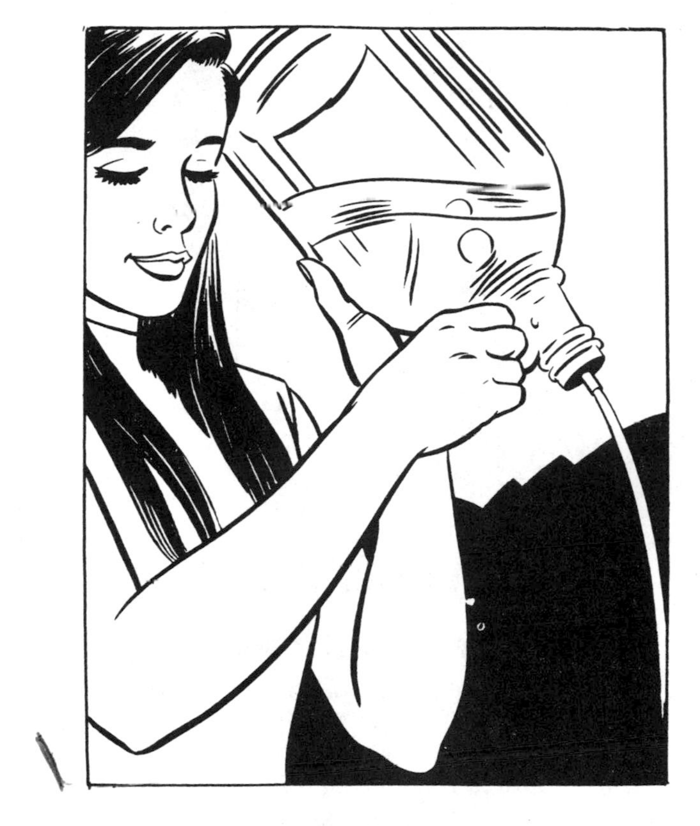

THE GURGLING JUG

Needed: A gallon jug, a short soda straw (large size) sealed into the cap with wax.

Do This: Fill the jug one-quarter full with water. Screw the cap on tightly and hold the jug in the position shown. The water makes a gurgling sound as it comes out in spurts.

Here's Why: When the jug is inverted, the slight downward motion of it will force out a spurt of water. If held still, an air bubble will creep up and allow the flow to start.

The flowing water decreases the pressure of the air above it in the jug, but the water's inertia keeps it flowing briefly, until the air pressure in the bottle is low enough to stop the flow and suck some water and air back into the jug. This upward flow increases the pressure in the jug, and another spurt of water is forced out.

The sequence is repeated over and over, and a gurgling sound is heard each time the water stops coming out and a bubble of air enters the jug.

3

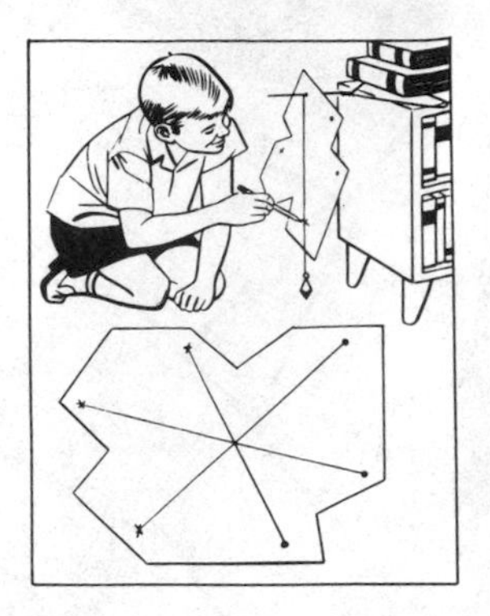

Mechanics

TO SEE WEIGHTLESSNESS

Needed: Coffee can with lid, a nail, string, metal-cutting shears.

Do This: Make a hole in the lid of the can. Run the string through it and tie the end to the nail. Place the can on the table, hold the nail up by the string, and let the nail fall. It falls into the can with a noise.

Next, hold the nail and can up by the string, and let go of the string so the nail and can fall together. The can will hit the floor first, then the nail. The nail and can have both been weightless as they fell, and the nail did not fall to the bottom of the can while both were falling together.

Here's Why: When both can and nail are released at the same time they tend to fall at the same rate (air resistance actually would reduce the speed of fall of the can slightly but in the experiment this can be ignored). An orbiting space capsule may be compared with the can, and its occupants are represented by the nail. The capsule is in continuous free fall, but its fall is so synchronized with its speed that it does not reach the earth. Its occupants are under similar conditions, which are described as weightless. The occupants do not tend to move in any direction in relation to the capsule.

By cutting as hole in the side of the can we can see the weightlessness of the nail as it and the can fall.

FLAME IN A JAR

Needed: A candle, large glass jar, strong cord, cardboard.

Do This: Attach the candle to a small square of cardboard with melted wax. Fasten the cord around the mouth of the jar so it cannot pull loose. Put the candle into the jar and light it. Then, holding the jar by the cord, turn around. The flame will point toward you as you turn.

Here's Why: The gases in the flame are hot and lighter than air. As you turn, the heavier air in the jar is thrown away from you by centrifugal force, so that the lighter flame is drawn toward you.

In the same way, a piece of wood floating on water in the jar will stay toward you as you turn with the container, because centrifugal force tends to make the heavier water go away from you. See top illustration, page 36.

SPECIFIC GRAVITY

Needed: A heavy object, string or cord, a spring scale, water, pencil and paper.

Do This: Weigh the object (it must be a solid object, not porous) in air. Then weigh it while it is immersed in water. Its specific gravity will be its weight divided by its loss of weight in water.

Here's Why: Specific gravity is the weight of the object divided by the weight of an equal volume of water. The difference in the object's weight in air and its weight under water equals the weight of the water it displaces, or an equal volume of water.

The coal in the drawing weighed 6 1/2 pounds in air and 1 1/2 pounds in water. The difference is 5 pounds. Divide 6 1/2 by 5 and the specific gravity of the coal is found to be 1.3. See bottom illustration, page 36.

HARDENING AND SOFTENING OF COPPER

Needed: Large copper wire, hammer, heavy metal to hammer on, alcohol lamp or gas flame.

Do This: Flatten a part of the wire by hammering it. It will be hardened where hammered. This is cold working of a metal. Heat it, cool slowly, and it will become soft again. This is annealing of a metal.

Here's Why: The copper is made up of grains or crystals with imperfections called dislocations. The dislocations can spread when the metal is bent or caused to spread by hammering. The metal is "soft."

But when the metal is made to bend or work continually, the dislocations reach a point where they resist being made to move further. This increases the force necessary to bend the metal. Then we say it is "hard."

STRAIN HARDENING

Needed: Copper wire, a nail in the wall, gas or alcohol flame.

Do This: Bend the wire (No. 12 or larger) and notice that it bends easily. Now pull the wire over the nail back and forth as if the wire were rope and the nail a pulley. The wire gets more difficult to bend.

Here's Why: Defects or irregularities in the ordered pattern of the copper crystals that make up the wire act like sand under a sled runner to impede the sliding of the crystal planes over each other. When the defects are in the middle of a crystal, they slide with the planes and offer only a little extra resistance. But when a defect reaches a grain or crystal boundary, it locks or jams the crystal plane on each side of it.

Heat the wire red hot, immediately cool it in water, and it becomes soft and ductile again because the defects will have diffused back into the crystals.

WAVE TRANSMISSION

Needed: Marbles, a grooved ruler.

Do This: Place five or six marbles in the groove so that they touch each other. Roll another marble against the end of the line of marbles. The vibration wave will be transmitted through the line and the marble on the end will roll away. Roll two against the line and two will roll away.

Here's Why: Waves in solids, liquids, and gases consist of a certain amount of matter in motion at a certain velocity. The waves move through the matter by transmitting the same amount of velocity to the same amount of new matter. Momentum, which is mass times velocity, is transmitted. One unit of momentum coming in makes one unit go out, and two units of momentum make two units go out. The same applies to three or more. The marbles must touch each other on the ruler. This is an example of conservation of momentum.

CRACK THE WALNUT

Needed: English walnuts.

Do This: Hold a walnut in the hand and try to crack it by a simple squeeze. It is difficult or impossible. Squeeze two together in the hand and one of them is cracked easily.

Here's Why: When one is squeezed, all the force exerted by the hand is distributed over a large area of the shell as the hand takes the shape of the shell. When two are squeezed together, a

strong force is exerted on small areas of shell where they touch.

Pressure is total force divided by the area involved. When two walnuts are used the area of contact is small, and the force divided by the small area is greater then the same force divided by the larger area of the hand. This is shown in drawings A and B.

INTERNAL STRESSES

Needed: Old glass telephone pole insulators, a 400-degree oven, long tongs, cold water.

Do This: Heat the insulator in the oven for one hour. Remove with the tongs and dip into cold water.

What Happens: Contraction of the outer surface of the glass produces such violent internal stresses that the insulator becomes a mass of tiny cracks. It has a crinkled or "crystalline" appearance.

NOTE: I prepared several dozen of these insulators in testing the safety of this experiment. There was no indication of flying glass or other danger. But it is best to wear goggles or at least turn the face away when the insulators are immersed in the water.

"CREEP"

Needed: Wire solder, copper or aluminum wire, weights, a measuring stick.

Do This: Hang the solder and the other wires (two or more feet of them) and suspend equal weights from them. Measure the distance of the weights from the floor, then after 24 hours measure again. The solder will have become permanently longer. The other wire probably will have become longer, but much less so.

Here's Why: Stretch something and let it go. It will go back to its original shape. This is elastic behavior. Stretch it again and again, each time a little more, and a point will be reached where it does not return to its original length or shape. You have exceeded the "elastic limit."

Stretch the wire below its elastic limit, but leave it stretched a long time, and it does not go back to its original length. This is "creep." Creep is flow and takes time as well as stretch. Creep occurs below the elastic limit but is faster the closer the stretch is to the elastic limit.

THE LARIAT

Needed: A rope and the skill to twirl it.

Do This: Twirl the rope as a cowboy does and the loop can be made to remain parallel to the ground.

Here's Why: When the rope is twirled each small section of it acts as an individual mass and tends to fly off on a tangent to get as far from the center of rotation as possible.

This is an example of centripetal force, which is the force acting toward the center of rotation and keeping the body moving in a circular path. The mass tends to move away at a tangent. The centripetal force keeps it from doing so.

When the rope is twirled rapidly, this force is sufficiently strong to overcome the pull of gravity which would pull the rope downward.

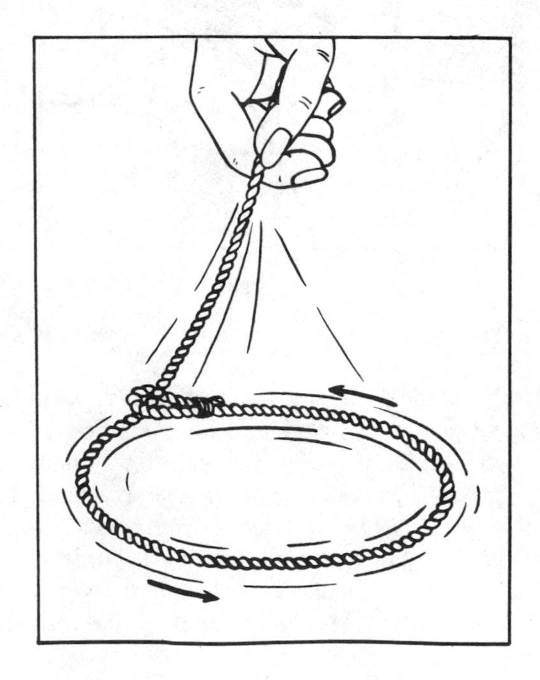

CENTRIFUGAL AND CENTRIPETAL FORCES

Needed: A small fishbowl, a marble or small ball.

Do This: Place the ball in the bowl. Hold the bowl upright and move it in a circle by spinning around on your feet like a ballet dancer as shown. The ball will climb the side of the bowl toward the outside of the circle.

Here's Why: The ball wants to move in a straight line and would do so if the side of the bowl did not constantly push it toward the center of your circle (centripetal force). Of course, the ball pushes back (centrifugal force) against the side of the bowl and so climbs to the widest part of the bowl, the part farthest from the center of your circle.

ROPE FRICTION

Needed: A rope and a tree.

Do This: Wrap the rope around the tree one time, hold one end of it, and have three people try to pull it out of your hand by pulling the other end. It is impossible.

Here's Why: Friction of the rope as it slides against the tree is very great; a possible figure is that if one end is held with a pull of one pound, nine pounds of pull on the other end will not move the rope. The frictional force becomes greater as the pull on the rope becomes greater.

NOTE: This is a problem that cannot be solved in general terms because the rubbing surfaces vary too much. It is possible that one pound of pull can equal 81 pounds on the other end of the rope.

THE ROLLING CANS

Needed: Two coffee cans, four plastic covers for them, eight large nails or spikes.

Do This: Cut the bottoms out of the cans and place the plastic tops on both ends of the cans. Push four nails through the plastic lids of each can, placing them close to the center in one and near the rim in the other.

Release the cans at the same elevation on an incline and allow them to roll. The can with the nails near the center will gain speed faster and will outrun the other. After the cans roll on the level floor, the slower one may catch up with and roll farther than the faster can.

Here's Why: The resistance to change in motion is known as rotational inertia. In this experiment, the rotational inertia of the

can with the nails near the rim is greater than the can with the nails near the center; therefore the can with the greater rotational inertia starts slower and is also more difficult to stop.

INERTIA

Needed: Two bottles, a card, a piece of writing paper, a five-cent piece.

Do This: Place card and coin on the bottle and by a quick flip of the finger the card can be moved from under the coin, allowing the coin to remain on the bottle top.

Place the bottles one atop the other, with the paper between. After a little practice it will be possible to jerk the paper away and leave the bottles standing. Jerk the paper with both hands as shown, or hold the paper with one hand and strike the center of the paper downward with the edge of the other hand.

Here's Why: Inertia in this case is the tendency of the coin or the bottle to remain in the same place unless moved by force. A force is applied to the coin and bottle by motion of the card or paper, but the force is not great enough to move the upper object if the paper or card is moved quickly.

NOTE: Bottle necks, paper, and card must be dry.

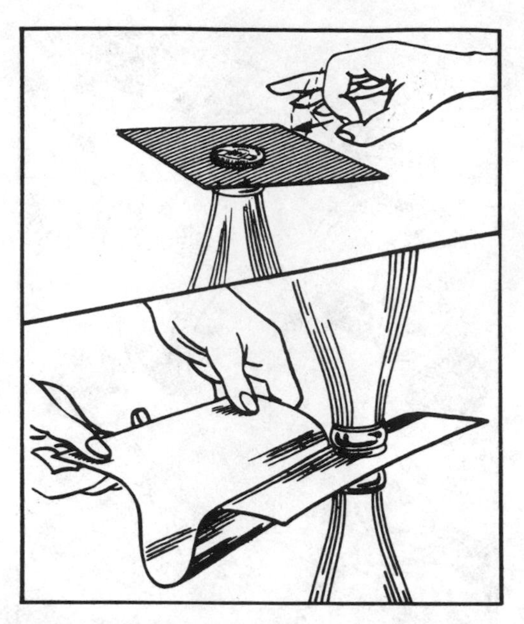

STATIC VERSUS MOVING FRICTION

Needed: Several long rubber bands and a weight such as a book.
Do This: Tie the bands together and attach them to the book.

Lay the book on a table or wood floor and pull on the bands with gradually increasing force. Notice that the bands stretch farther (we pull harder) to start the book moving than to keep it moving after it is started.

Here's Why: Static friction is greater than moving friction for most solid surfaces. Thus, it takes more force to start the motion than to maintain it. This does not apply to liquid friction, which has no tendency to stick to a surface. For example, a touch of a finger can move a boat on water.

FIND THE CENTER OF GRAVITY

Needed: Cardboard, scissors, string, a weight, pencil, coat hanger.

Do This: Cut a piece of cardboard in an irregular shape. Place the hanger on a shelf so that a wire extends out. Punch three or four holes in the card at various edges. Hang the card on the wire and hang the string on the wire in front of the card. The weight is used to keep the string straight.

Make a cross at the lower end of the card behind the string. Hang the card by the other holes, and each time make a cross at the bottom of the card.

Draw a line to connect the holes and corresponding crosses, and it will be seen that all the lines intersect at the same point. This point is the center of gravity of the card, the point at which it will balance on the end of a wire.

BEAM STRUCTURE

Needed: Two yardsticks, clamps, tacks (or glue), string and weight, chairs for support.

Do This: Place the yardsticks one on the other between the chairs. Place the weight on their middle and measure the bending. Then clamp, glue, or nail the sticks together so they cannot slide one over the other. Note that they do not bend as much as the weight is applied.

Here's Why: In this type of bending the upper surface of each stick is compressed or made shorter while the lower surface is stretched or made longer. With the sticks merely placed one on the other this change in length can be made easily by one surface sliding over the other.

If the sticks are held rigidly together, the surfaces cannot slide one over the other. It is almost as if we had one stick twice as thick. One stick twice as thick will bend only one-eighth as much, other factors remaining the same.

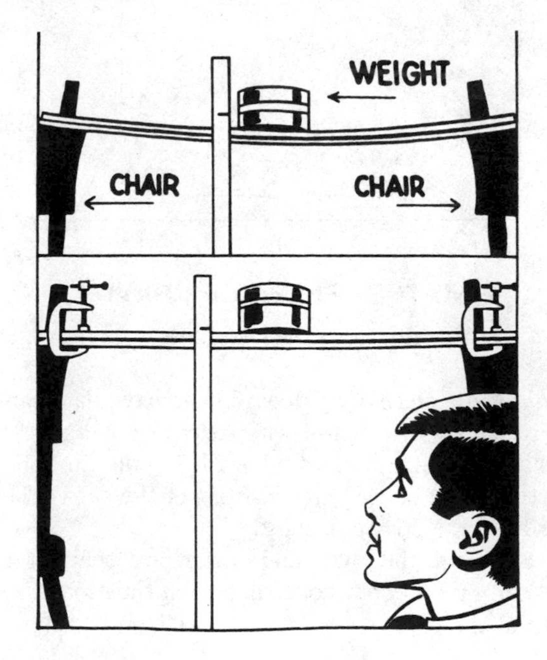

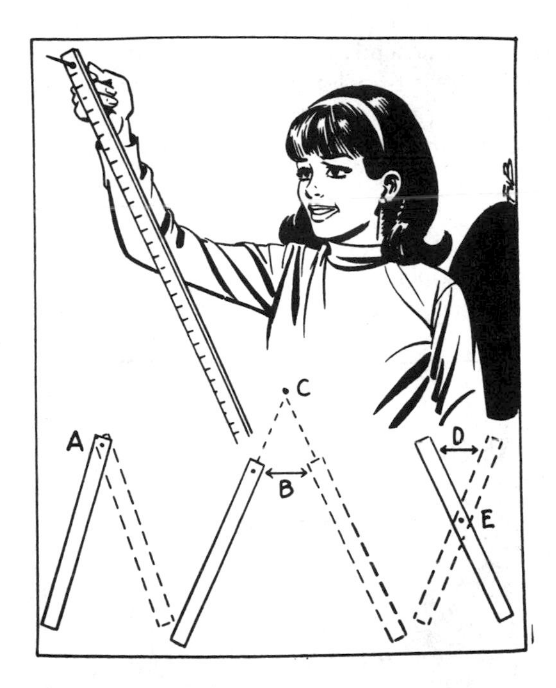

THE CRAZY PENDULUM

Needed: A yardstick with a hole in one end, a nail.

Do This: Hold the yardstick on the nail, hold the hand still, and start the stick swinging as a pendulum. The stick will have a normal swinging time or "period" and the pivot point is at A.

Now, move the hand back and forth, more slowly than the normal period of the stick, and the pivot point will seem to be above the stick at C. The upper end of the stick will move as at B. If the hand is moved back and forth faster than the natural period of the stick pendulum, the pivot point will move down to below the end of the stick, and the swing will be as shown at D and E.

Here's Why: The law of the pendulum says that a free, unforced pendulum has a period proportional to the square root of the length. When we force the pendulum to swing at a rate different from that, it virtually adjusts its length by moving its pivot point so that it still can follow its law.

THE SYMPATHETIC PENDULUMS

Needed: Two weights, string, a way to suspend the string.

Do This: Tie the weights C and D to the string as shown. The long string A-B may be suspended in a doorway. Start one pendulum swinging. The second pendulum will begin to swing, and soon most of the energy is transferred to the second one. The first will almost stop. Then the action will be reversed, the second pendulum gradually losing energy as the first one swings farther and farther.

Here's Why: Energy can be neither created nor destroyed. As the second pendulum begins to swing, it is taking energy from the first, until the first pendulum has lost about all of its energy. The energy transfer begins to take place in the opposite direction.

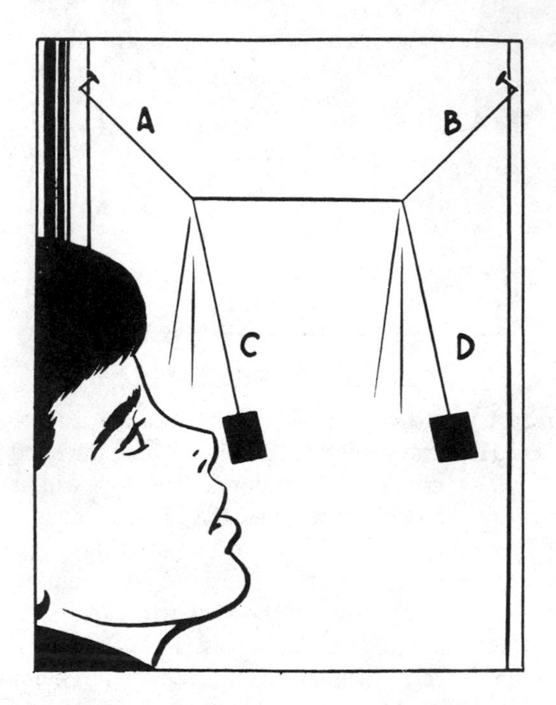

HOW BALL BEARINGS WORK

Needed: Two similar paint cans, some marbles.

Do This: Place the marbles around the rim of one of the cans, and place the other can down over them as shown, so that the marbles will remain in the depressions of the cans near the rims. The top can will turn very easily on the marble ball bearings.

Here's Why: Ball bearings reduce friction in automobiles and other moving machines. Friction caused by rolling is much less than sliding friction, because the balls are hard and make very little contact with the surface over which they roll.

CONSERVATION OF ENERGY

Needed: A strong string, a thread spool, a light weight.
Do This: Thread the string through the spool and tie one end

to the weight. Holding the other end of the string in one hand and the spool in the other hand, swing the weight around in a wide circle over the head. Do this out of doors and away from anything breakable.

Pull down on the string, pulling the weight toward the spool, and the weight will turn around faster and faster.

Here's Why: When the weight is started around in a large circle it is traveling at a certain speed. As the radius of the circle is shortened, the weight tends to move at the same speed so that there is no loss of energy. But since the distance around the circle is less, the weight must make more revolutions per minute to move with the same speed.

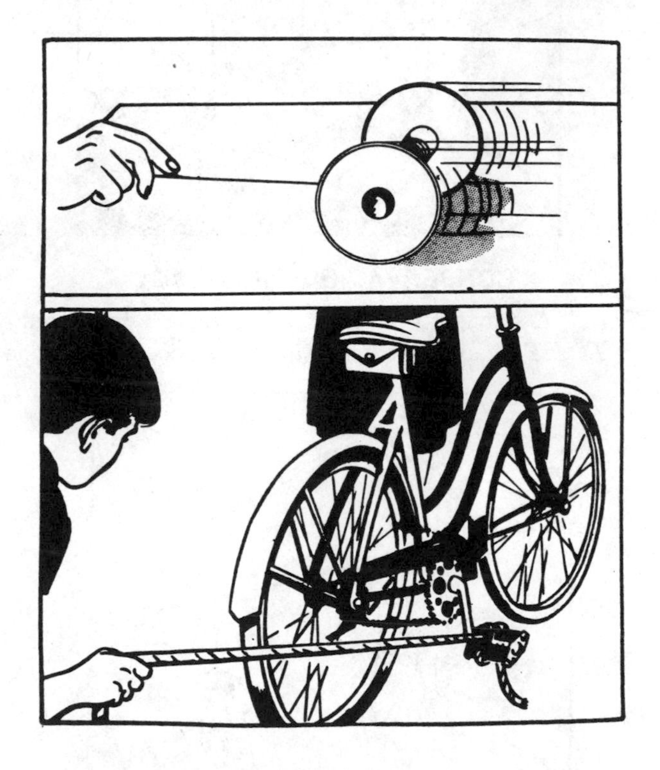

THE BICYCLE

Needed: A bicycle, rope and two people.

Do This: Tie the rope to a pedal and turn the pedal down. Have someone hold the bicycle so it does not fall over. Pull backward on the rope and the bicycle moves backward.

Here's Why: Although the bicycle moves backward as measured against the ground, the pedal will move forward as measured against the bicycle frame.

The sprocket ratio and wheel diameters are such that when the end of the pedal moves an inch, the bicycle moves several inches. Thus, when you pull backward on the pedal both the bicycle and pedal move backward, as does the hand, as measured by the ground.

When you pull on something, your hand will never move in the direction opposite to the pull unless the thing pulled has a motor in it.

The principle involved can be seen on a tabletop or floor. Pull on a thread wound around a spool and the spool will roll toward the hand, winding up the thread. Note that here, too, the hand moves in the direction of the pull.

4

Projects to Build

MAKE A BASS FIDDLE

Needed: An old-time wash tub, a stick, strong string, tools.

Do This: Cut a notch in the end of the stick so it will stay on the rim of the tub. Make a small hole in the center of the tub, put the string through, and tie a knot so it will not pull out. Tie the other end of the string to the stick as shown. Place your foot on

the rim of the tub to hold it down. You can make music by plucking the string.

Here's Why: When the string is plucked (a D string from a bass viol is good for this), it vibrates. As it transfers its vibrations to the tub, the tub also vibrates, and its sound is loud enough to be heard clearly.

As the string is tightened, it vibrates faster and so does the bottom of the tub. With practice, musical notes may be produced.

NOTE: A five-gallon paint bucket may be used instead of the tub, and strong cord such as fishing staging may be used instead of the D string. If a piece of 2-by-4 wood is placed under the edge of the tub below where the notched stick rests, there can be an improvement in the sound. It gives an outlet for the vibrations coming from the bottom side of the tub base.

A SURVIVAL STILL

Needed: A plastic sheet, container, stones, a digging tool.

Do This: Dig a hole in the ground and place some leaves or other green plant materials along the sides of the hole. Place the container in the middle and drape the plastic as shown in the drawing. Distilled water will drip into the container when the sun shines.

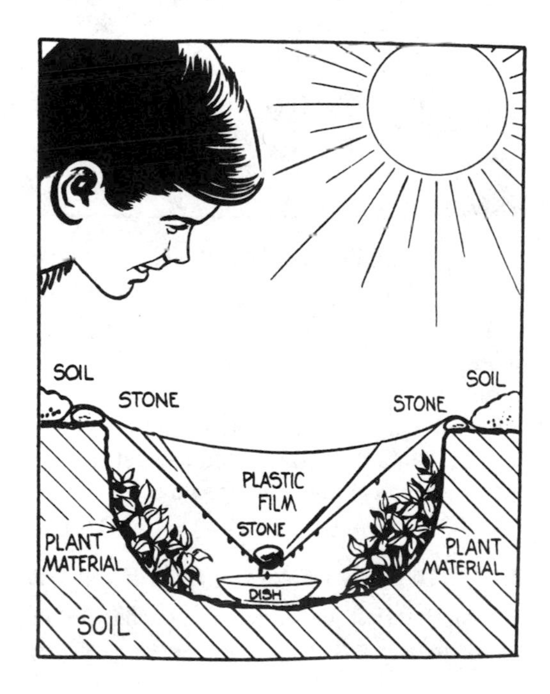

Here's Why: Sunlight passes through the plastic and is absorbed by the earth and plant materials. This produces warmth which in turn causes evaporation of water, most of which collects in droplets on the underside of the plastic. The water runs down and drips into the container.

NOTE: Ray D. Jackson and C. H. M. van Bavel presented the idea for this still in the magazine *Science*, suggesting that it could be used to obtain water for survival in the desert.

A SIMPLE TURBINE

Needed: A tin can or milk carton, string, nail, hammer, water.

Do This: Attach the container to a string, make holes in it as shown, suspend it, put water in it, and it will begin to turn as the water spurts out.

Here's Why: Sir Isaac Newton found that every action has an equal and opposite reaction. As the water is expelled from the holes, it pushes on the container in the opposite direction, and this is what makes it turn.

ILLUSTRATING NEWTON'S THIRD LAW

Needed: Smooth level balsa wood boards, smaller blocks, dowels, cloth for a sail, a thread spool, screws, a large balloon.

Do This: With screws attach the spool to an upright block. Bore a quarter-inch hole in the block for an air passage. (The block is attached to another block as a base.) Attach the sail to another block bearing upright dowels.

Place one board on dowels over another, so the rolling motion is free. Place the sail and balloon on the upper board so the air blows against the sail. Nothing happens.

Lift the sail and the board rolls along on the dowels in the direction of the arrows. Replace the sail, lift the balloon, and the board with sail moves in the other direction.

Here's Why: When both sail and balloon are on the board, the reactions are equal and opposite between them. When the sail is lifted, the reaction as the air escapes freely from the balloon moves the board on which the balloon rests. When the balloon is lifted so the air blows against the sail, it simply pushes the board along as wind pushes a ship when it blows against the ship sails.

COMMENT: Newton's Third Law states that for every action there is an equal and opposite reaction.

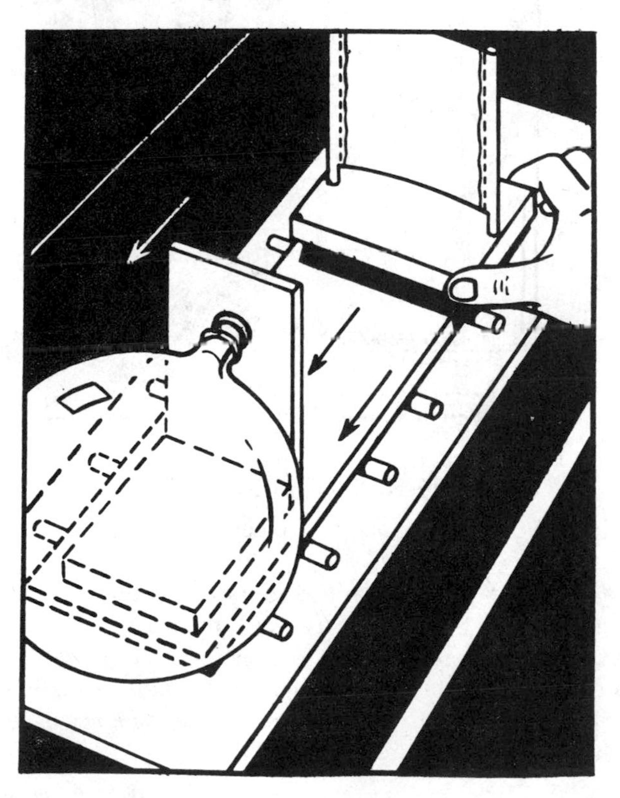

A TROLLEY RIDE

Needed: A strong wire, two pulleys, wood, bolts.

Do This: The dimensions are not too important, but examples are shown in the diagrams. Construct as illustrated. Attach the wire between two trees and you have a very fine ride for young children. The greater the slope of the wire the faster this trolley will run. Do not let the trolley and human load bump dangerously against the supporting tree at the lower end. Put some protective padding at the lower end, or make the slope of the wire so slight that the trolley will stop before it reaches the lower end.

THE HAIR HYGROMETER

Needed: A wooden support, pin, pencil, glue, cardboard, a few strands of long human hair.

Do This: Set up the apparatus as shown, so that the cardboard pointer is held up by the hairs (two or three hairs are plenty). The hair expands as the moisture in the air increases, causing the pointer to go down. This type of hygrometer gives approximate relative humidity. A wet-and-dry bulb hygrometer is more accurate and is used by the weather bureau.

LIGHT UNDER WATER

Needed: Enameled magnet wire, finishing nails, cardboard tube from bathroom tissues, bulb for a one-cell flashlight, jar of water, toy train transformer.

Do This: Cut the tube the length of the nails, stuff nails into it and wind about 200 turns of wire around it, securing it with tape. Wind 50 turns around a shorter piece of the tube and solder the ends of the small coil to the bulb. Connect the larger coil to the transformer.

Place the coil and bulb in the jar of water, bring the jar down over the larger coil, and the bulb should light under water.

Here's Why: Water can conduct high voltages and kill but will not short out the low voltage from the transformer. The two coils here make a transformer. They are connected to each other by a magnetic field that is effective through the water and the glass of the jar.

NOTE: Dr. Elihu Thomson, an American scientist, made his coil and bulb light in weight so they would barely sink in the water. The electric repulsion made them rise when the current was turned on. This is difficult but would make an excellent science fair project if worked out successfully.

A BIMETALLIC STRIP

Needed: Coffee can, 14-gauge copper wire, soldering iron, wood.

Do This: Cut two strips of metal from the can and solder them together end to end. They should be a little more than two feet long. Wind the metal around a small wooden cylinder (which may be cut from a large dowel) and attach it firmly with screws. Solder the clean bare wire along the middle of the strip.

Take the assembly off the wood, and attach one end of it to a support, leaving the other end free as in the lower drawing. This makes a crude thermometer.

Here's Why: With changes in temperature the copper expands and contracts more than the iron of the tin can does. This is the principle of most thermostats and some thermometers.

NOTE: I cut my metal strips one-half inch in width, but the measurements are not critical.

MAKE A NIGHT LIGHT

Needed: Old glass telephone pole insulators, a wood base, lamp cord, a socket such as is used in some Christmas tree light strings.

Do This: Assemble the light as shown in the drawing. If the socket is not tight, put glue around it. It is advisable to put felt on the bottom of the base to prevent scratching furniture on which the lamp may be placed.

Christmas tree bulbs of different colors may be used and clear or green insulators. The insulators may be difficult to find in some places. An employee of the telephone company may be able to supply one.

If the insulator is "crystallized" as explained earlier, the night light is more attractive and unusual.

THE SUN'S ECLIPSE AND CORONA

Needed: Wood, cardboard, black paint, wire, dowel, electric bulb in socket with cord attached, tools.

Do This: Assemble as shown. The dowel slides through a hole in the middle block so that it can pull the wood frame holding the light back and forth. The small holes should be about one-sixteenth of an inch in diameter; they can be made with a sharp pencil.

Before the experiment begins, adjustments must be made so that the round cardboard "moon" exactly hides the hole in the large card through which the bulb shines. In my model the "moon" is 1 1/4 inches in diameter, and the "sun" hole is 2 inches. The

"moon" is on a wire and may be adjusted up and down to align it with the holes and the "sun."

What to See: Begin by looking through one of the holes near the edge of the card and the bright "sun" can be seen. As you look through holes nearer the center of the card the eclipse appears. When looking through the center hole the "moon" hides the sun, but the "corona" may be seen.

The "corona" in this case is due to diffraction of light in the atmosphere around the moon. The center hole, "moon," and "sun" must be line for this.

A SOLAR STILL

Needed: An empty rectangular gallon can, tin snips, sheet of window glass, two clothes pins, clay, metal kitchen foil, water.

Do This: Slice the can open as shown with the tin snips. Put clay around the edges to act as a seal, attach the clothes pins to the top of the glass so that it will not slide, bend the foil into a trough, and you have a solar still.

How It Works: When water is placed in the still and bright sunlight is directed into it, the water evaporates and much of it condenses again on the undersurface of the glass. It runs down the undersurface of the glass, dripping into the trough and then into a container.

The still will work in bright sunlight or under a strong lamp. It is usually necessary to paint the inside of the still black. This will change more of the entering light energy into heat and so evaporate more water.

The small still is not practical for producing usable amounts of distilled water, but the principle is shown, and several ounces of distilled water may be obtained in this manner. It is better to start it with hot water.

5

Tricks

THE FALLING BALL

Needed: A 2-by-4 timber, three or four feet long, two small cups, a ball bearing or other iron weight, glue or tacks.

Do This: Fasten the cups at the end of the timber a few inches

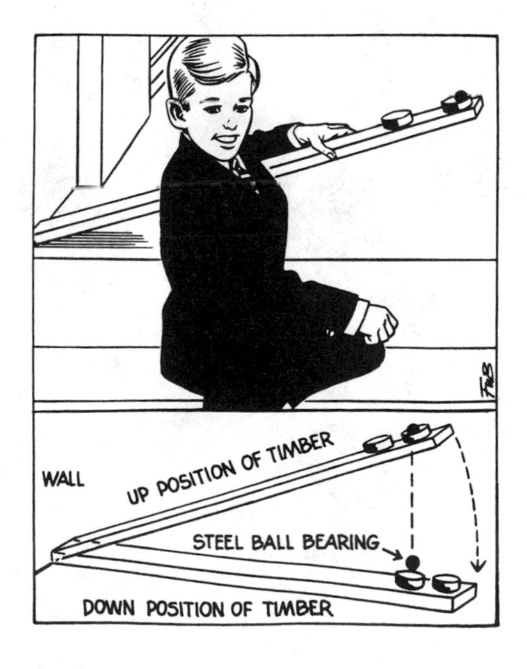

apart. Place the weight in the cup nearest the end, raise the timber, and let it drop. After some practice the ball may be made to transfer to the other cup as it falls. (Place the lower end of the timber against the wall so it will not slip as it falls.)

Here's Why: A freely falling body gains 32 feet in speed every second it falls because of the constant pull of gravity. But the timber is not a freely falling body. The end in contact with the floor does not fall at all. So, the point three-quarters of the way to the free end falls as if it were free. The upper end falls faster than free fall, because it is pulled down not only by gravity but by the rest of the timber.

The ball, being free, cannot fall faster than 32 feet per second. So, the cup moving faster than that runs away and leaves the ball, which falls straight down. The cup follows the arc of a circle, allowing the second cup to fall below the falling ball.

BREAK THE MATCH STICK

Needed: Wooden kitchen matches.

Do This: Place the match in the fingers as shown and try to break it with finger muscles. It is difficult or impossible. Hold the center finger on a table and the stick may be broken by pushing

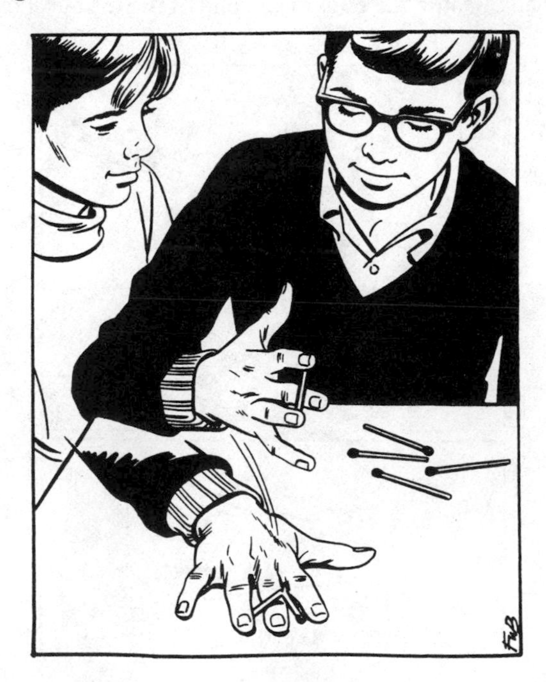

with the adjacent fingers, or hit the hand on the table and the stick breaks.

Here's Why: The muscles of the hand are so placed that there is little strength to force the fingers backward. The table top furnishes the force necessary for the middle finger to hold itself while the other fingers break the match.

To be safe, strike each match carefully and blow out the flames before trying this trick. Then a match head cannot ignite and burn a finger.

THE DISC GYRO

Needed: An old long-playing phonograph record, string, match stick.

Do This: Thread the string through the hole in the record. Tie the stick to one end of it, so that the record rests on the match stick when you hold the string out by the other end.

Challenge someone to swing the record back and forth as a pendulum, keeping it level. Ask that it not be allowed to flop around. The friend may find it impossible.

Then give the record a twirl, and as it swings back and forth, it remains level, or in the same plane.

Here's Why: This is a gyroscope, the machine that guides airplanes and rockets. The spinning top is another gyroscope. The law is: every turning body tends to turn in the same plane in which it begins to turn, unless some force outside moves it from that plane.

HIDE THE PENNIES

Needed: Seven coins, a glass of water, a plate.

Do This: Arrange the coins as shown, place the glass of water over them and put the plate over the glass. Look down toward the coins as in the drawing, and they seem to disappear, except for the edges that are seen around the outside edge of the glass.

Here's Why: It is a matter of refraction, which is the bending of light rays when they pass at an angle from one medium to another, in this case air, glass, and water. Rays from the coins under the glass are bent upward so that they are cut off by the plate and never reach the eye.

INERTIA TRICK

Needed: Smooth table, smooth wrapping paper, a glass or other container of warm water. The container must have a smooth bottom.

Do This: Dry the outside of the glass thoroughly and place it on the paper as shown. Strike the paper quickly with one hand while holding it with the other and it will slip from under the glass leaving the water unspilled.

Here's Why: If the paper is pulled slowly, the glass is pulled with it because of the friction between the paper and the glass. But if the paper is pulled quickly, the inertia of the glass and water is great enough so that the frictional force does not cause much motion of the glass.

Use warm water because cold water may cause condensation on the outside of the glass, wetting it and increasing the friction so that the glass may be pulled off the table.

THE COIN ON EDGE

Needed: A large coin.

Do This: Try to stand the coin on its edge. It is difficult. Spin it and it stays on edge until it slows down. To spin the coin, hold it vertically and snap the first finger of the right hand against the outer edge of the coin.

Here's Why: The coin, as it turns, becomes a small gyroscope which resists any change in its plane of rotation. This is one of the simplest gyroscopes. They were invented in 1852 by a great French

physicist, Leon Foucault.

A bicycle stays upright partly because its spinning wheels become gyroscopes. Complicated gyroscopes, called gyrocompasses, guide ships and most airplanes.

A CHINESE PUZZLE

Needed: Four or six square pieces of thin plywood or thick cardboard three or more inches square, wide twill tape from a store, glue or a staple gun.

Do This: Glue or staple the blocks together with the tape as shown in the drawing, leaving about a half-inch of space between them. Hold by the top block, invert it, and one block seems to fall through the others to the bottom.

Here's Why: The blocks are hinged with the tape so that one is allowed to tumble downward after the other. It is a small demonstration of "chain reaction" and is quite amusing.

BALANCE THE BOOKS

Needed: About ten books of the same size, a sturdy table.

Do This: Make a straight pile of the books, then move the top

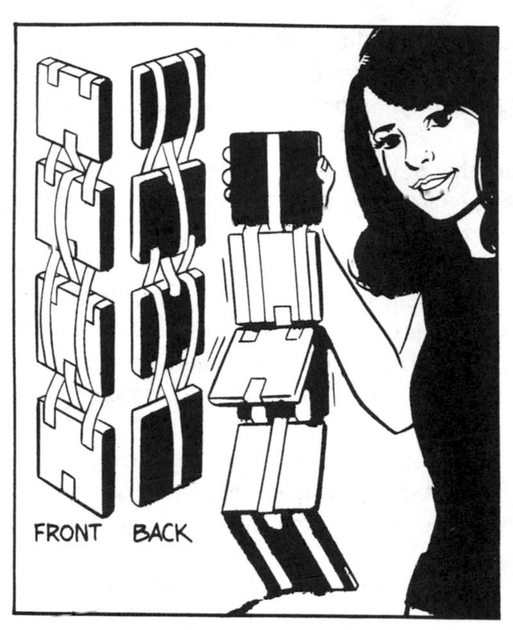
FRONT
BACK

book outwards until it almost topples. Move the second from the top. You will find that it can be moved almost a quarter of its length over the edge of the third book before it and the top book are likely to topple.

Continue this moving down the pile, and when the bottom is reached it will be found that the top book extends entirely over the edge of the table.

Here's Why: When the first book is moved, it can be moved almost halfway over the edge of the second because its center of gravity is in its center. As the second book is moved the center of gravity of the two top books combined must be considered. So, it can be moved almost a quarter of the way over the edge of the third.

As the third book is moved, the center of gravity of all three books determines the extent of overhang that can be achieved. Each book may be moved a little, but the overhang must be a little less each time because the center of gravity to be considered must be that of the book being moved and all above it.

INVISIBLE WRITING

Needed: One drop of cooking oil, a tablespoon of strong household ammonia, water, small brush, writing paper, small bottle.

Do This: Put the oil and ammonia into the bottle. Add about four tablespoons of water and shake thoroughly. Using the brush, write on the paper with this solution.

When the paper is dry the writing disappears, but it will appear again when the paper is dipped in water. It may be made to appear and disappear many times. If the liquid is used again, it must be well shaken.

Here's Why: Oil and water do not mix, but the ammonia helps dissolve the oil so that when the mixture is shaken the ingredients are fairly well mixed. When they are painted on the paper the oil is absorbed, while the water and ammonia are evaporated. When the paper is dipped into water, the water is not absorbed as much where the oil is, so there is enough contrast to make the writing visible.

SECRET WRITING

Needed: Milk, paper, a toothpick or small brush, an electric iron.

Do This: Dip the toothpick into the milk and write on the paper with it. Let the paper dry and the writing will have almost completely disappeared. Heat the iron on the wool setting, pass it over the paper, and the writing will appear.

To avoid any possible damage to the iron, place another sheet of paper over the one containing the writing so the iron does not actually touch the milk.

Here's Why: The dried milk chars more easily than the paper and so shows up as darker lines.

THE CLOCK AS A COMPASS

Needed: Sunlight, a clock or watch, a straight stick (a match will do).

Do This: Hold the clock face horizontally; hold the stick vertically so that its shadow falls along the hour hand. South will be halfway between the hour hand and the 12.

Here's Why: The sun is the same relative position at a given time during any season of the year, relative to the horizontal clock face.

The time given by this method is sun time or standard; it will not be accurate for daylight-saving time. In winter, when the sun is lower in the sky, the stick would have to be placed at the edge of the dial so that its shadow would fall along the hour hand.

HEAT CONDUCTION

Needed: A tall, slender juice can, water, electric immersion heater, wire mesh, ice cubes.

Do This: Hold the can of water by the bottom. Hold the heater element in the center of the can, just under the water. Do not let the handle of the heater get wet. The water will boil at the top of the can, but remain cool enough to hold at the bottom. Next, cut a piece of wire mesh so it fits snugly in the can as shown. Place ice in the bottom of the can, and insert the mesh to hold the cubes at the bottom of the water. The water in the upper part of the can will boil, yet the ice cubes in the bottom will not melt.

Here's Why: As the water is heated it expands, becomes lighter, and floats on top of the cooler water. Conduction of heat downward through the water is very slow, but convection currents set up by the warming carry some warmer water to the lower parts of the can.

A STRENGTH TEST

Needed: A piece of pipe 3/8-inch in diameter and a little shorter than a pencil, a pencil, paper, rubber band, some sand or salt.

Do This: Put a piece of paper over the end of the pipe and secure it with the rubber band. It may be pushed off easily by pushing the pencil through the pipe. Do the same, except this time put sand or salt into the tube so that it fills the tube about one-third full. It will probably be impossible to push the paper off by push-

ing against the sand.

Here's Why: In the second instance the force exerted by the pencil is largely converted into sidewise forces which push against the side of the pipe. This causes friction between the grains of sand and the side of the pipe so the push on the paper is small.

INSIDE THE LIGHT BULB

Needed: Two electric bulbs, one lighted; cardboard, large pin.

Do This: Make a large pinhole in the cardboard. Hold it against the lighted bulb. Hold the unlighted frosted bulb against it and the shadow of the wires inside the second bulb will be projected clearly on the glass opposite the pinhole.

Here's Why: Light travels in straight lines through the small hole in the opaque card. The wires in the unlighted bulb interfere with the direct flow of light and cast a shadow.

NOTE: This is the principle of the pinhole camera. See top illustration, page 77.

A FEATHER MYSTERY

Needed: Feather with a white end, two pieces of cardboard, rubber bands, pencil, bright light.

Do This: Using the pencil, make holes in the cardboards at ex-

actly the same point. Place the feather over one hole and cover with the second card. Fasten cards with the rubber bands. Facing a bright light, look through the holes and through the feather at the same time. Colored patterns may be seen, and if looking at the hand an "x-ray" picture apparently shows the bones in the fingers.

Here's Why: White light is wave motion with waves of many different lengths. As the light waves pass through narrow slits such as the feather, they are broken up and act as if starting at each slit anew.

The waves meet again at the eye but their path lengths now are different so that their crests and troughs may or may not coincide. Where wave meets trough there is no light.

Since the colors have different wave lengths, this happens at different places for different colors, and so the white light is broken up, making color bands.

The light acts as if bent at the slits, so that some of the light which would have been stopped at the edges of the fingers reaches the eye, making the finger edges look partly transparent. The centers of the fingers look solid because light from that region is not bent enough to get through.

Thus, a solid core is seen surrounded by a semitransparent rim. The solid core may resemble the bones in the finger. See bottom illustration, page 77.

A RUBBER BAND TRICK

Needed: A yardstick, rubber bands, tacks or small nails, a wire, a lamp in reflector.

Do This: Find a four-drawer cabinet or dresser with knobs. Loop the rubber bands around the top knob and attach to the yardstick as shown. Attach the wire to the yardstick, and tie it around the bottom knob so that the tension keeps the yardstick horizontal.

Apply heat to the rubber bands from the lamp and they do not expand—they contract. The end of the yardstick may be seen to move upward. (The right end of the stick is free to move.)

Here's Why: Rubber is one of the few substances that contracts rather than expands when heated. Rubber also heats when stretched and cools when allowed to return to its normal state. This can be felt by stretching a flat rubber band quickly while it is held against the lips (Le Chatelier principle).

A TRICK OF THE EYES

Needed: A card, pin, and a light source.

Do This: Make a pinhole in the card and holding the card at arm's length look at the light through it. Bring the pin near the eye and into the line of sight between the hole and the eye. The pin will appear upside down.

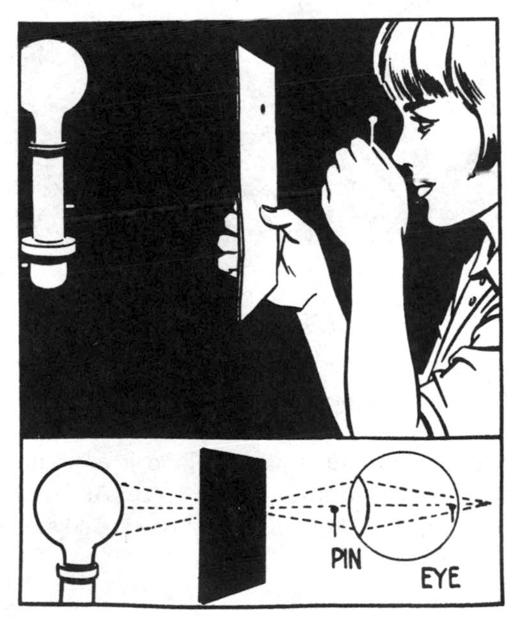

Here's Why: The simplified diagram shows the paths taken by the light rays from the source to the eye. The pin in this case is a shadow cast on the retina of the eye. Since the shadow is right side up the brain will see it as being upside down. This may not be a simple experiment to do on the first try.

NOTE: Several scientists to whom I have shown this experiment have questioned it. Anyone who does not agree with the diagram and explanation is invited to write to me and express another opinion.

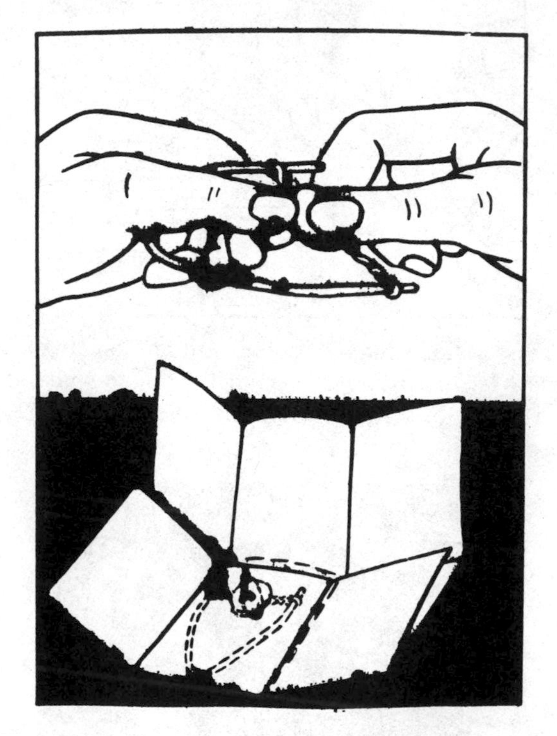

"RATTLESNAKE'S TEETH"

Needed: Button, rubber band, stiff wire, paper.

Do This: Thread the band through the button, attach the ends of the band to the wire, twist the button rather tightly, and place the device carefully in folded paper. Hand it to an unsuspecting friend and tell him it is a rattlesnake's teeth. As he unwraps the paper, the button revolves rapidly, making a sound somewhat like that of a rattlesnake.

Here's Why: This is an example of energy. As the button is twisted, the potential energy is built up in the rubber band. It remains until the button is allowed to turn, in which case the energy becomes kinetic energy, which is energy of motion.

A MYSTERY COLOR WHEEL

Needed: White cardboard, string, pencil, black paint, a red light, protractor and ruler.

Do This: Draw and cut out a cardboard circle. Paint half of it black and cut a notch in the edge of the black side as shown. Put the disk on the string so it can be twirled by the twisting and untwisting of the string. Then look at a red light through the notch. Various colors can be seen.

Here's Why: Receptors in the retina of the eye are each more sensitive to one color than to another, but they are confused when a color is not continuous. They are fooled into believing they are seeing colors other than the one that reaches them intermittently.

NOTE: I admit this is not a good explanation, but as far as I know the has never been a good one. Anyone who wishes to comment is invited to do so.

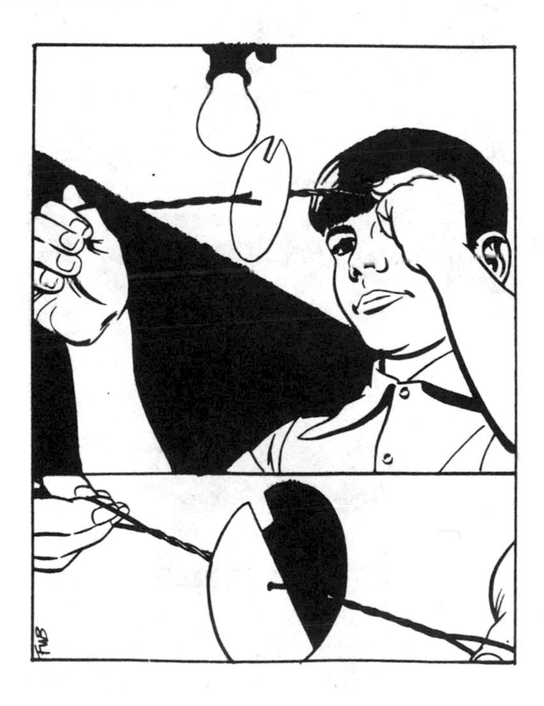

MAKE A TRICK BOTTLE

Needed: A plastic dishwashing detergent bottle, string.

Do This: Put the string through the hole in the cap and tie a large knot to keep it from sliding out (the string must slide very loosely in the hole). Hold the string by the cap and lower it into the bottle. Replace the cap. By squeezing the bottle quickly the string is forced out, looking like white liquid detergent. It's quite a surprise to the person on whom the "liquid" is squirted.

Here's Why: As the bottle is squeezed, air rushes out through the hole very quickly, dragging the string with it. Also the air pressure inside the bottle, which is greater than that outside, tends to force the string out.

COMMENT: This experiment was suggested by Terri Fulks, age ten, Oak Ridge, Tennessee.

NOTE: Use only a soft string, and it will be harmless even if it hits the face. Do not use any liquid in the bottle.

A TOUCH PUZZLE

Needed: One glass marble.

Do This: Place the marble in the left hand. Cross the middle finger of the right hand over the pointer finger. Close your eyes

and touch the marble with the tips of the crossed fingers. Roll the marble around with the fingertips; it will feel like two marbles.

Here's Why: Our senses are conditioned by habit and our brains interpret their signals in the manner to which they are accustomed. The feel of the marble on what would normally be opposite sides of the two fingers is interpreted as two marbles.

COMMENT: This experiment was suggested by Paul DeYoung, age ten, Sioux City, Iowa.

6

Chemistry

STARCH

Needed: Cornstarch, water, a pan, heat.

Do This: Try to mix the starch with water. It does not dissolve completely. Boil the water and starch mixture and it becomes thick.

Here's Why: Cornstarch and other starches are composed of grains or granules. In hot water the grains burst and dissolve into a thick paste. They do not burst in cold water.

NOTE: Starch itself is not a food, but when eaten it is broken down by enzymes to form various sugars and monosaccharides, which are energy foods. If the starch is heated to 350 degrees Fahrenheit or hotter, it is changed into dextrin, a substance halfway between starch and sugar. Dextrin can dissolve in cold water and is supposedly the sticky substance used on postage stamps.

The heat of toasting changes some of the starch in bread into dextrin, the dark surfaces of a toasted slice. The enzyme ptyalin, present in the saliva in the mouth, starts the digestion of starch by changing some of it to dextrin.

PREPARE OXYGEN

Needed: Jar, dry yeast, hydrogen peroxide (6% or 20% volume) from the drug store, matches, wooden splint.

Do This: Place a spoonful of yeast in the bottle and add about three spoonfuls of peroxide. A bubbling will be seen as oxygen is released from the peroxide. Place a glowing splint into the bottle and it should flame brightly. This is the test for oxygen.

Here's Why: The yeast is a catalyst, a substance that can speed up a chemical change without itself being changed. As the hydrogen peroxide (H_2O_2) loses some of its oxygen it becomes water (H_2O).

COMMENT: This experiment was suggested by Dr. Francis W. McCarthy of Boston State College, writing in *Science and Children*. He credits the original idea to Dr. Elbert C. Weaver of Phillips Andover Academy, who demonstrated it at a meeting of the American Chemical Society.

CARBON

Needed: A candle and a piece of metal.

Do This: Hold the metal (a can lid will do) in the flame and it is soon covered with soot.

Here's Why: The flame heats the wax of the candle, producing from it carbon, water, and several other substances. When the candle burns undisturbed, the carbon unites with oxygen from the air, producing colorless carbon dioxide. But if the flame is cooled by the metal, much of the unburned carbon, which cannot unite with oxygen at a reduced temperature, is deposited on the metal as soot.

NOTE: The girl is holding the metal with a piece of folded cardboard to prevent burning her fingers.

SHINY ALUMINUM

Needed: An aluminum pot turned dark (the dark stain is usually aluminum sulfide), a slice of lemon, water.

Do This: Boil water in the pot and drop in a slice of lemon. The pot will brighten. Acid foods such as tomatoes boiled in the pot will brighten it also.

Here's Why: Aluminum is the second most active of the common metals; only magnesium is more active. Either of these metals would react violently with water or air, except for a film of oxide that forms on the surface and protects it from further action.

Citric acid from the lemon combines with aluminum oxide to form aluminum citrate which dissolves in the water, leaving the bright metal exposed. Other aluminum compounds are formed in this way when fruits are cooked in the aluminum and eaten—most people say safely—with the food. The dark stain comes away with the oxide.

NOTE: At least one scientist, while believing aluminum compounds are harmless to the body, does not use aluminum utensils because "no one knows what aluminum as a powerful reducing agent does to food elements such as vitamins."

A LOT OF CARBON

Needed: Gum camphor from a drug store, metal can lid, a match, can of cold water.

Do This: Put a pea-sized lump of camphor on the lid, light it, and hold the can of water above it so the soot collects on the bottom of the can. This soot is almost pure carbon and can be scraped off in powder form. This is lampblack, a common paint pigment.

Here's Why: The camphor contains many substances, carbon being the least easily oxidizable. When the flame is cooled by the can, the carbon is below the temperature of its ready oxidation, and the free carbon is deposited there.

NOTE: Add a little linseed oil to the lampblack to make a paste, then add turpentine, and you have a flat black paint.

THE GAS FLAME

Needed: A gas stove, an electric stove or hot plate, a pan or kettle of cold water.

Do This: Place the pan or kettle over the gas flame, and note that it collects moisture on the outside; it does not when placed over the hot electric unit.

Here's Why: The gas flame, in burning, produces mainly water and carbon dioxide, along with a few other substances in very small quantities. The water it produces is in the form of steam which condenses on the cold metal. The electric burner does not produce either water or gases (unless food or other substances have been spilled on it).

PHYSICAL CHANGE OF A COMPOUND

Needed: Sugar, warm water, a dish.

Do This: Put sugar into a quarter cup of warm or hot water. Add sugar until no more will dissolve. Let the undissolved sugar settle. Then pour the solution into a dish. Set the dish aside until the water evaporates.

What Happens: The water evaporates; the sugar does not. The sugar is left behind, on the dish, but while it is the same in chemical composition as the crystals used in the beginning, it is in a different physical form.

JUICE AS A CLEANER

Needed: Juice from apples, plums, or other fruit; a copper-bottomed pan or a strip of dirty copper.

Do This: Prepare the juice by boiling the fruit or peelings in water, as if to make jelly. Place the pan or the strip in the juice, leave it several hours and the copper will be much cleaner and brighter.

Here's Why: Fruit juices contain a variety of diluted plant acids which react very slowly to dissolve or separate copper oxides, hydroxides, or basic carbonates (dirty copper). This exposes the free copper underneath the surface film and restores the original copper color.

If only part of the copper is placed in the juice, the contrast between the cleaner and the dirty copper will be seen. In the drawing a spoon is being placed to hold one side of the pan out of the juice.

YEAST

Needed: One cup water, two teaspoons sugar, one-quarter teaspoon powdered dry yeast, one cup flour, two bowls.

Do This: Put a half cup of water into each bowl. Add one teaspoon sugar and a half cup of flour to each. Put the yeast in only one bowl. Mix the ingredients and place the bowls in a warm place.

In an hour or so the bowl in which the yeast was placed will contain a bubbly mass, while the other remains almost unchanged. The yeast, which consists of tiny plants, feeds on the sugar, and gives off alcohol and carbon dioxide.

Here's Why: The growing yeast plants produce an organic catalyst which causes the chemical change. If this process takes place in bread, the carbon dioxide gas bubbles make the bread rise and the alcohol is boiled off in the baking.

REMOVE SCORCH

Needed: A slightly scorched cotton cloth, hydrogen peroxide such as is found in hair bleach.

Do This: Moisten the scorched place with the peroxide, then place another cloth over it. Iron it dry. If the scorch is not all gone, repeat the process.

Here's Why: In this experiment hydrogen peroxide (H_2O_2) in contact with the scorched cloth becomes plain water (H_2O) as it loses some of its oxygen. The lost oxygen atoms are very active, looking for something to combine with. We call them "nascent" atoms. They combine with the light brown of the scorch to form a colorless product.

This will not work on a badly scorched spot where the cloth fibers have been severely damaged.

SOAP ACTION

Needed: Olive or cooking oil, water in a bottle, soap shreds or detergent.

Do This: Put a few drops of oil into the bottle of water and shake. When the water is still, the oil comes to the top. Put a little soap or detergent into the bottle, shake again, and the oil and water will mix. They do not separate when the water is still.

Here's Why: Soap and detergents have long molecules, one end of which will dissolve in water and the other in oil. Shaking breaks up the oil into tiny droplets. The oily end of the soap molecules dissolves in the droplet, leaving the watery end sticking out into the water.

Thus each oil droplet is coated with a "water-liking layer" so when two droplets bump together they do not combine or merge into one because each is protected by its repellent film. They remain small, spread through the water more or less evenly. This is an "emulsion."

NOTE: Soap is a detergent. When the word "detergent" is used here, it refers to one of the synthetic detergents.

HARD WATER

Needed: Two glasses half full of water, one teaspoon Epsom salts, one-half glass of soapy water.

Do This: Add the salts to one glass of water. Put half of the soapy water into each glass and stir. Note that the glass without the salts makes suds; the other does not. A scum forms in it.

Here's Why: The magnesium ions in the salts combine with the soap to form insoluble soap which in turn forms a froth, scum, or precipitate.

Soap is a detergent, and the effectiveness of any detergent depends on its concentration in the water. Combining the magnesium with the soap removes the soap from the water.

Synthetic detergents are more effective than soaps in cleaning because they do not form insoluble compounds so readily.

MAKE GLUE

Needed: A pint of skim milk, half pint of vinegar, enamel or stainless steel pan, a bowl, one teaspoon baking soda.

Do This: Put the milk into the pan, add the vinegar, heat and stir until lumps form. Pour the lumpy mass into the bowl. When it cools, pour off the clear liquid on the top.

Dissolve the soda in a quarter cup of water, add it to the lumpy

mass and a chemical action takes place producing casein glue.

Here's Why: Acetic acid in the vinegar coagulates the casein of the milk, making lumps which may unite to form one large lump. The baking soda, a base, neutralizes the acid of the vinegar and disperses the coagulated casein to form a smooth gluelike fluid.

SUPERSATURATION

Needed: Crystals of photographer's hypo, water, a pan, heat, a straight-sided glass, a ruler, a clean drinking glass.

Do This: Mark the straight-sided glass with five lines equally spaced from each other, beginning at the bottom of the glass. This is the measuring vessel. Put in five measures of hypo, dump it into the pan, measure one-fifth as much water, put it into the pan, and heat the mixture gradually until all the hypo is dissolved.

Pour the clean liquid into the glass and let it cool. It will remain liquid. Now put in one crystal of hypo, and watch. The crystals will begin to grow in a most beautiful manner, until the entire liquid will be changed into one mass of crystals.

Here's Why: The liquid when cool is supersaturated; that is, it contains more of the dissolved hypo than it normally can hold. The forces are so delicately balanced that it requires a slight dis-

turbance, such as jarring or the addition of another crystal to start the crystallization process, which, once started, continues until a stable equilibrium is reached.

RUST

Needed: Three clean nails, three dishes, oil, water, sand and salt.

Do This: Rub a thin film of oil on each nail and place them in the dishes. Put plain water in one dish, sand and water in another, and salt and water in the third. Set the dishes aside.

In a few hours, certainly overnight, the nail in the salt water will show signs of rust. The other nails will begin to rust, but later.

Here's Why: Salt solution partly destroys the oil film which protects the nails against rapid rusting. Whenever impure iron is in contact with water, oxygen and carbon dioxide, rust will begin to form—oxygen combines with the iron to form iron oxide.

The sand has no effect on the rusting process unless it rubs some of the oil film off, thus starting the rusting sooner.

SAILOR'S SOAP

Needed: A bowl, water, salt, soap, salad oil.

Do This: Put water into the bowl, put oil on a hand, and wash

the hand in the water, using soap. The oil washes off. Repeat, but first put two heaping tablespoons of salt into the bowl of water, and stir until the salt is dissolved. The oil cannot be washed off the hand with the soap.

Here's Why: Soap, which is soluble in fresh water, is not soluble in salt water because of the "common ion effect." Soap is a sodium salt of a fatty acid. Salt is sodium chloride. The common ion, sodium, reduces the solubility of the soap. Salt-water soap, the kind sailors use at sea, contains potassium instead of sodium, and is not rendered insoluble by the sodium in the ocean water.

IODINE STAIN

Needed: A few ounces of water, a few drops of household tincture of iodine, a few crystals of photographer's hypo.

Do This: Stain the water with the iodine. Drop a few crystals of hypo into the solution and it becomes clear. Try removing an iodine stain from a piece of cloth by putting a strong solution of hypo and water on it.

Here's Why: A chemical change takes place. The hypo, which is sodium thiosulphate, unites with the brown or purplish-brown iodine stain to form a colorless compound called sodium tetrathionate. Only a definite amount of hypo can unite with a definite amount of iodine. Use enough.

CHICKEN BONE CALCIUM

Needed: Two chicken bones (or other bones), two glasses, water, vinegar.

Do This: Place a bone in each glass. Cover one with water, the other with vinegar. Pour out the water and vinegar two or three times a week and replace with fresh.

In two or three weeks the bone in the vinegar will be flexible; the other will be unaffected by the water and will remain stiff.

Here's Why: Bones are hard and firm due chiefly to calcium phosphate which is not soluble in water, but is slowly changed by the acetic acid of vinegar into soluble calcium acetate; so as the bone loses its calcium phosphate it loses its stiffness and becomes flexible.

This experiment does not suggest that vinegar taken into the stomach is harmful to the bones. Ingested, it can never reach the bones as acid vinegar.

MAKING BUTTER

Needed: A quart of whole raw milk, a bowl, a half-gallon jug, a spoon—and plenty of patience.

Do This: Let the milk set in the bowl in a warm place 24 to

36 hours. Put it into the jug and shake until the fat has changed to butter. This takes 15 minutes or more. Wash out the milk with water and butter is left.

Here's Why: The fat in milk is not in solution but is in invisible droplets called an emulsion. The shaking or churning causes the particles to cling together to form little balls of butter which can be seen.

Harmless varieties of bacteria such as *Bacillus bulgaricus* grow in the milk as it sets in a warm place and produce lactic acid. This is what makes buttermilk sour. The acid increases the tendency of the milk fat to aggregate.

NOTE: The increase in size of the fat particles—butter making—is not a chemical change.

A TEST FOR VITAMIN C

Needed: Cornstarch, water, tincture of iodine, orange juice, a heat source, containers.

Do This: Boil a teaspoon of cornstarch in a cup of water. This dissolves some of the starch. Put ten drops of this mixture and one drop of tincture of iodine into half a glass of water. Add food containing vitamin C, such as orange juice, drop by drop, until the blue color disappears. Try fresh orange juice, then test some that has been boiled for five minutes. Note the difference.

Here's Why: Starch and free iodine unite to form a substance of unknown composition called starch-iodine. The delicate blue-purple color is removed by heating or adding enough fruit juices containing vitamin C.

Boiling citrus fruit destroys vitamin C, in part.

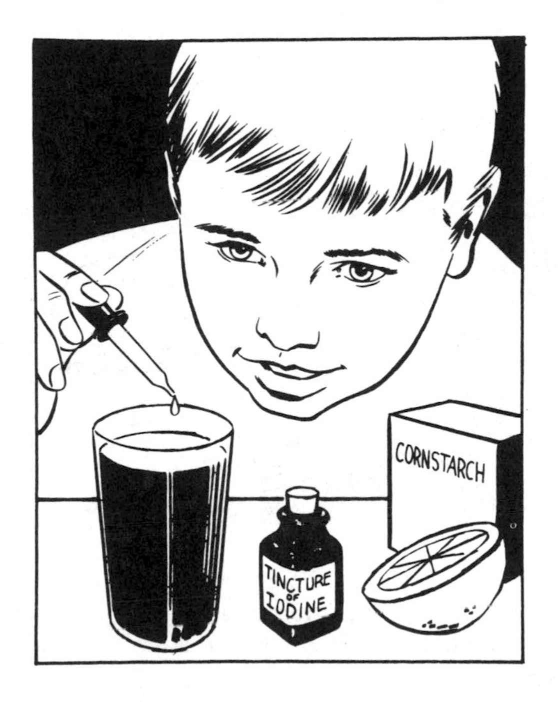

THE MEANING OF "ANHYDROUS"

Needed: Magnesium sulphate crystals (commonly called Epsom salts), a tin can lid, heat.

Do This: Heat some of the crystals on the lid. Water will be given off in the form of steam, and the crystals will turn into a white powder.

Here's Why: Epsom salt, like the sulfate salts of several other metals, can form beautiful crystals that are even more beautiful under a microscope. The shape is the result of combining one molecule of the salt with a certain number of molecules of water. Heat boils away the water, leaving the powdery salt behind.

The crystals are called hydrate crystals, from the Greek word *hydor*, meaning water. When the letter *alpha* is placed before a

Greek-derived word, it has the effect of "not" or "without." So a hydrate crystal becomes an anhydrous powder as it loses its water.

NOTE: Hydrate is a noun. Anhydrous is an adjective. Dehydrate is a verb.

CORROSION

Needed: Three jars with tight-fitting lids, three large iron nails, water, a stove, a large container, sandpaper or emery cloth.

Do This: Clean the nails thoroughly with sandpaper or emery cloth. Boil two of the jars in deep water in the large container. Put nails in each and boil some more. Seal one jar with its nail. Leave the other open. Remove them from the boiling water. Place cold tap water in the third jar and put a nail into it.

The nail in the unboiled water and that in the open jar of boiled water rust in a few hours. The nail in the sealed jar of boiled water resists rust for many hours.

Here's Why: Corrosion or rusting is caused not by water alone,

but by the oxygen and carbon dioxide dissolved in the water also. Boiling removes most of the dissolved gases from the water. The water in the open jar reabsorbs air; the water in the closed jar reabsorbs practically no air.

"LIVE" RUBBER

Needed: Balloon rubber, calcium carbide, water.

Do This: Place a piece of the rubber on a newspaper and put a lump of calcium carbide on it. Pour a little water on the chemical. The rubber will squirm as if alive.

Here's Why: When the water touches the calcium carbide, a chemical action begins, changing the mixture into water-slaked lime and acetylene gas, producing heat. The heat causes the rubber to contract unevenly, and the contraction causes the squirm.

NOTE: Calcium carbide is sold in hardware stores for use in lamps and in toy stores for use in toy cannons. If it is fresh it may have a protective coating on it that prevents quick reaction with water. In this case, put a drop or two of water on the carbide to remove the coating before trying the experiment. See bottom illustration, page 103.

7

Light

FLYING SAUCERS?

Needed: A plastic shoe box, water, milk, flashlight, opaque tape, a dark room.

Do This: Place tape over the face of the flashlight so a quarter-

inch slit remains for the light to pass through. Put four or five drops of milk in the box and fill with water. Stir. Shine the light as shown and it will reflect downward from the surface of the water.

Here's Why: Light striking the surface of the water is reflected downward. In the May, 1970 issue of *Popular Science* magazine, an article by David Heiserman suggests this phenomenon as a possible explanation for some flying saucer sightings. Light, he points out, can reflect from an atmospheric temperature inversion, which is a situation where warm air rests above cooler air. Automobile headlights from a car going uphill can reflect down on the other side of the hill miles away, and a small turning of the car can make the reflected lights move very fast.

EASY DIFFRACTION INTERFERENCE PATTERNS

Needed: Two fingers and a light.

Do This: Hold your fingers near your eye. Look at the light between the fingers with one eye and slowly bring the fingers together until they almost touch.

What Happens: As the fingers are brought closer together a dark structure will suddenly appear to jump out from each finger to make a connection between them. Examine carefully. Each finger has alternate light and dark bands near it, like waves on a beach. The dark structure is the intersection of these two systems of bands.

Here's Why: Light is wave energy. Water waves striking a breakwater bend a little into the calm space behind the breakwater. In the same way light waves striking the fingers bend a little into the dark space behind the fingers. When two such light patterns overlap, they alternately reinforce and cancel each other. This is called interference. See top illustration, page 107.

HOW HIGH IS A TREE?

Needed: A tree, a straight stick, a sunny day.

Do This: To find the height of the tree in the well-known manner, stand the stick up and measure its height. Also measure the tree's shadow and the stick's shadow. Use inches for all the measurements. Multiply the length of tree's shadow by the height of the stick, then divide by the length of the shadow of the stick.

COMMENT: This method is correct as long as the tree has a pointed top. But if the tree has a round shape, this method cannot be depended on because the shadow observed may be the shadow of the side of the tree, not the top. See bottom illustration, page 107.

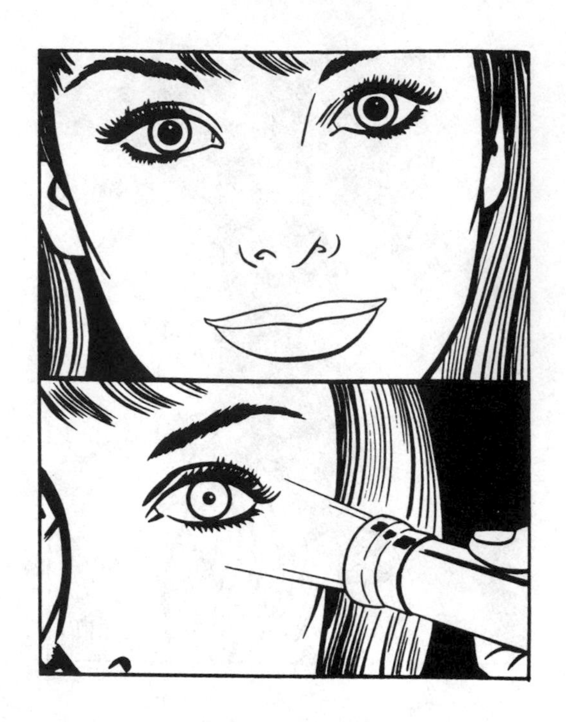

THE SENSITIVE PUPIL

Needed: A flash light.

Do This: Notice the size of the pupil in a friend's eye. Then shine the light into the eye and watch the pupil get smaller. When the light is removed the pupil will gradually go back to its normal size.

Here's Why: The eye sees better if the right amount of light comes into it. The pupil is the valve by which the eye adjusts itself so that the right amount of light is allowed to shine in. In dim light the pupil is large; in bright light it is small. Various disorders of the body can cause the pupil to stop working properly.

REFRACTION OF LIGHT

Needed: A deep glass dish, flashlight and support, mirror, water, milk, talcum powder and puff, cardboard with a slit in it.

Do This: Set up the experiment as shown so that the flashlight beam goes through the slit and into the water at an angle. A drop or two of milk in the water and dust in the air from the powder puff will scatter some of the light so that its path can be seen bending.

Here's Why: The speed of light varies in different substances; it is greater in air than in water. When the beam hits the water at an angle the side nearest the water is slowed down first, the side farther from the water last. This causes the beam to bend like a sled when one runner hits sand while the other is on snow. When the beam comes out of the water after reflecting from the mirror, it bends the other way.

NOTE: The mirror is placed in the bottom of the dish, under the water.

THE MOON ILLUSION (1): WHY DOES THE FULL MOON LOOK LARGER WHEN IT FIRST RISES?

Needed: A piece of glass and a candle.

Do This: Smoke the glass until the candle flame can be seen through it, but not the surrounding objects. Look through the glass at the sun or moon at the horizon and it will appear the same size as when it is high in the sky. (Do not look at the sun, even through a dark glass, when it is bright. Do this experiment only as the sun is rising at the horizon.)

Here's Why: The moon at the horizon appears larger because the surrounding objects and terrain give the impression that it is

closer and larger than it is.

NOTE: This explanation was advanced in 1693 by the French philosopher Malebranche, and was quoted in the September, 1962 issue of *Science* by Norman I. Harway of the University of Rochester.

THE MOON ILLUSION(2)

Needed: A yardstick, paper clip, moonlight.

Do This: Bend a paper clip as shown so it will stay on the end of a yardstick. With the moon at the horizon, bend the ends of the clip so they barely enclose the image of the moon when sighted from the other end of the yardstick. Later in the evening look through the clip again. You will see that the ends of the clip still barely enclose the moon's image.

Here's Why: The apparent larger size of the moon at the horizon is an optical illusion. See illustration, page 111.

A LIGHT MYSTERY

Needed: Eggshell, candle, glass of water.

Do This: Hold a piece of eggshell in the candle flame until it

MOON
BENT
PAPER CLIP
END OF
YARDSTICK
FWB

FWB

is smoked black. Then look at it under water. It will be a silvery color.

Here's Why: The flame deposits lampblack and a little cracked paraffin on the eggshell. This mixture is hydrophobic—it hates water. Therefore, when the shell is immersed, the water does not wet it. It carries thousands of tiny air bubbles with it as it goes under the water.

Light reflects from this air—thus the silvery appearance. A spoon smoked with the candle flame will show the same reflection. Or, try putting a piece of black paper over the top of a glass in which there is water, and look up at the water surface. It will be silvery because of the reflection. See bottom illustration, page 111.

INTERFERENCE LINES(1)

Needed: Black paper, two razor blades, a fluorescent light.

Do This: Cut two parallel slits in the paper *very* close together. This can be done by cutting with both blades at the same time, holding them with thick paper between. Look at the light through both slits at the same time. Alternating light and dark lines can be seen.

Here's Why: This is one of the proofs of the wave nature of light.

Imagine water waves striking a solid object, represented by the opaque paper. Imagine how they would pass through the openings, then form other waves on the other side of the openings.

These seconds waves would overlap each other, in some places canceling each other out and in other places reinforcing each other. Thus, the light waves cancel out and reinforce each other, making the alternate dark and light lines that are seen.

INTERFERENCE LINES(2)

Needed: Two pieces of window glass, paper, rubber band, fluorescent light.

Do This: Put the pieces of glass together as shown, with glass touching at one end and the paper between the other ends. Hold them so the fluorescent light is reflected in them. Interference lines, suggesting fingerprints, can be seen. The lines move as the glass pieces are squeezed together.

Here's Why: There are actually four reflections of the light from the four surfaces of the glass pieces. The two inner surfaces are so close together that the reflected light waves tend to overlap at places reinforcing each other to make brighter lines and at other places canceling each other out.

NOTE: I used glass pieces 4 by 6 1/2 inches, with newspaper thickness at one end. The pieces were uneven enough to show the interference lines without the paper between them.

TV AS A STROBOSCOPE

Needed: A bicycle or other wheel with spokes, a TV set, a dark room.

Do This: Turn the TV brightness on fully and the contrast off, so the screen will be almost solid white. Turn the wheel. As it gains speed it appears at times to stand still or reverse its direction.

Here's Why: The TV screen does not remain constantly bright, but gets brighter and dimmer at regular intervals. The eye does not notice this because the image made on the retina of the eye takes a fraction of a second to fade out.

When the wheel turns so that a spoke is always at one spot when the light is brightest, and moves when the light is dimmest, the wheel seems to stand still. When the wheel turns so that a spoke is slightly before the spot, the wheel seems to creep forward, and if the spoke is slightly behind the spot, it seems to creep backward.

LIGHT AND DARK CLOUDS

Needed: Several sheets of tissue paper and a light.

Do This: Look at the light through the papers and the papers will appear dark. Hold the papers so that the light falls on them and reflects to the eye and they will look light.

Here's Why: Clouds are made up of tiny droplets of water, each of which acts as a lens to bend light rays and as a reflector to reflect them.

If the cloud is between us and the sun, and if it is thick enough, most of the light is reflected back toward the sun and the cloud looks dark. If the sun shines on the lower side, the reflections and refractions from the droplets come back to us and the cloud is white. The tissue paper reflects and refracts the light much as the cloud droplets do.

COLORS

Needed: A dark room, flashlight, colored cloth or paper, colored cellophane.

Do This: Shine the light on red cloth or paper and it will look red. Place the red cellophane over the flashlight and the red paper still looks red. Place blue cellophane over the light and the red paper will look black. Try other color combinations.

Here's Why: When white light shines on the red paper, red light is reflected while most of the other colors of light are absorbed.

Red light coming through the red filter on the flashlight still is reflected.

Almost no red light comes through the blue cellophane filter, and so the red paper looks black because there is no red light to reflect from it.

We observe colored objects by light transmitted through them or by light reflected from their surfaces.

THROUGH THE LOOKING GLASS

Needed: Pencil, paper, mirror.

Do This: Make a circle on the paper, place the mirror behind the paper, then try to trace around the circle with the pencil while watching your movements only through the mirror. It will be very difficult if not impossible to trace the circle in this way.

Here's Why: When you push the pencil away from yourself, the mirror image of the pencil is coming toward you. This confuses ordinary muscular movement with what you see in the mirror. With practice and patience we can learn to coordinate our muscle movements with the mirror image, as we do when we tie a necktie or put on makeup.

Another Mirror Stunt to Try: Print the name Bob Brown in capital letters, lay it on the table in front of the mirror, and the word Bob will appear to be normal, while Brown will be upside down.

THE WINDOW REFLECTION

Needed: A window.

Do This: Stand in front of a window on a bright day and the outside world is clearly seen. Stand at the same place at night, when lights are on in the room and the reflection of the room, not the outside, is seen.

Here's Why: A glass window pane reflects and transmits light. In daytime the bright light from objects outside comes through the glass, and while there is some reflection, the transmitted light is so much brighter that it hides most of the reflected light.

At night very little comes from outside through the glass pane. The much brighter light inside the room is reflected, hiding almost all the outside light that comes through.

The reverse is true from outside. Look in, and in daytime the outside light is reflected. At night the lighted room is clearly seen.

THE CONCAVE MIRROR(1)

Needed: A concave (magnifying) mirror, carbon paper, cardboard.

Do This: Attach a piece of carbon paper to the cardboard so it can be held still. Focus the sun's rays on it so that a small image of the sun is produced. The heat will melt the carbon.

Here's Why: Concave mirrors act much as lenses and are used in the largest telescopes. The light rays striking them are reflected as shown in the drawing to produce a hot spot. The area of the light spot is much less than the area of the mirror, but the small surface area receives most of the heat from the total surface of the mirror. See top illustration, page 119.

THE CONCAVE MIRROR(2)

Needed: A concave mirror, white cardboard, a room with windows on one side.

Do This: Hold the mirror and card as shown and move the mirror toward or away from the card. A distance will be found where an image of the window will show on the card.

THE SUN
HOT SUNLIGHT SPOT
MIRROR

A POINT ON
THE OBJECT
A POINT IN THE IMAGE
MIRROR

Here's Why: The rays of light from the window are reflected from the mirror as shown in the diagram. The mirror acts much as a lens to focus the rays of light. Light from the window is seen as a smaller image on the card. The light is reflected from an almost infinite number of seemingly flat points on the mirror, the angle of incidence equaling the angle of reflection.

The angle of incidence is that angle formed by the incoming ray of light with the perpendicular; the angle of reflection is the angle formed by the reflected ray with the perpendicular. See bottom illustration, page 119.

MIRACLE WITH LIGHT

Needed: Argyrol, water, an old spoon, sodium hypochlorite (household laundry bleach). Get argyrol from the drug store; tell the druggist you are using it for a science project so he will know it is not for medicinal purposes.

Do This: Put a half teaspoonful of argyrol into half a glass of water and mix. The solution should be dark brown. Add four teaspoons of bleach and the solution turns pearly white. Put the solution in bright sunlight and it will turn violet, then brown.

Here's Why: Chlorine from the bleach unites with silver in the argyrol to form a precipitate of silver chloride, which is silver-colored or white as the particles cling together. Sunlight decomposes the silver chloride to colloidal silver particles which are violet or brown, according to their size. This effect of sunlight in chemistry is still not well understood, although it is used in photography.

LIGHT BY FRICTION

Needed: Saran wrap and a fluorescent lighting tube (an old one will do).

Do This: In a dark room, rub the tube vigorously with the saran wrap. The tube will glow where it was rubbed.

Here's Why: A fluorescent tube glows when there is an electric field inside the glass. The field separates some electrons from their nuclei, ionizing the gas. Then as the electrons fall back into their regular places, the "ground state," they give off light.

The electric field is produced by the rubbing. In this case electrons are rubbed off from their atoms, producing positive and negative charges, and these charges can add up to a surprisingly high voltage. There is no danger in this experiment, however, unless the tube is broken. Then the danger is broken glass.

THE LIGHT PIPE

Needed: A glass jar with metal lid, nail to punch holes, cardboard, flashlight, small tube and rubber hose, water.

Do This: Punch two holes in the jar lid. Run water into the jar through one of the holes; let a stream run out the other hole. Hold the flashlight at the bottom of the jar and the light will be seen to follow the curving stream issuing from the hole. Cardboard may be wrapped around the jar and flashlight, but this is not necessary. A darkened room is required.

Here's Why: When light strikes a smooth transparent surface of a different medium it divides, a portion of it entering the new medium and a portion reflecting back. If the angle is steep more of the light goes through. But in the water stream, when the curve is less than 49 degrees, enough is reflected back and forth between the surfaces to reach the end of the stream where it touches the sink. A little light comes out at ripples along the stream making the stream glow. (The critical angle of 49 degrees applies to water; it is different for different media.)

CROOKED FINGER VISION

Needed: An index finger and printed matter.

Do This: Crook the finger as shown in the drawing, so there is only a small opening between the knuckles. Look through the opening with one eye and printed matter that may not be read with the naked eye may be read easily.

Here's Why: Light from an object normally hits the entire lens of the eye and much of it goes through, although the eye is often unable to focus the rays on the retina properly. When the light passes through a small area of the cornea and lens, as it does when coming from a small opening between the knuckles, the lens is able to adapt to the narrow light beam and this beam falls on the most sensitive part of the retina.

COMMENT: The Eskimos knew this trick before the first lens was made by Western man. They made glasses for their old people by drilling holes in whalebone and holding the bone on with leather thongs.

RAINBOWS IN A BUBBLE

Needed: A thread spool, a hose, bubble solution.

Do This: Blow a bubble and notice that bright rainbow colors may be seen in it. Notice also that the bubble appears black just before it's ready to burst.

Here's Why: When ordinary white light strikes the soap bubble, light is reflected from the front and back surfaces of the film. Due to irregular interference bright colors appears. The colors change as the thickness of the film changes during evaporation. As the thickness of the film becomes smaller than the wave length of ordinary light, there is destructive interference of all light and the bubble appears black. Some light passes through the film but the reflected light gives the colors.

NOTE: The diagram shows the reflections of a light ray from the two surfaces of a bubble. The drawing is greatly exaggerated.

A MIRROR TRICK

Needed: Two mirrors, two rubber bands, two books, a pencil.

Do This: Attach the mirrors to the books so that they stand upright. Place the pencil between them and many images of it can be seen.

Here's Why: The light reflected from the pencil reflects again many times from one mirror to the other before it reaches the eye. The diagram shows the paths of some of the light rays.

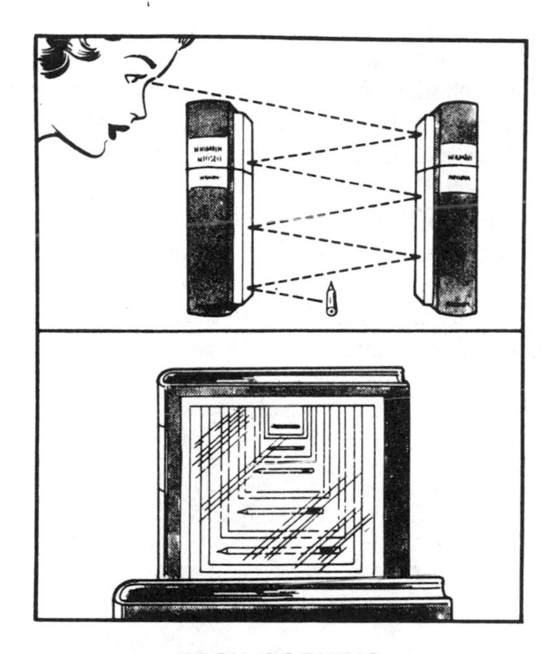

EASY COPYING

Needed: A pane of glass held upright, a picture and drawing paper.

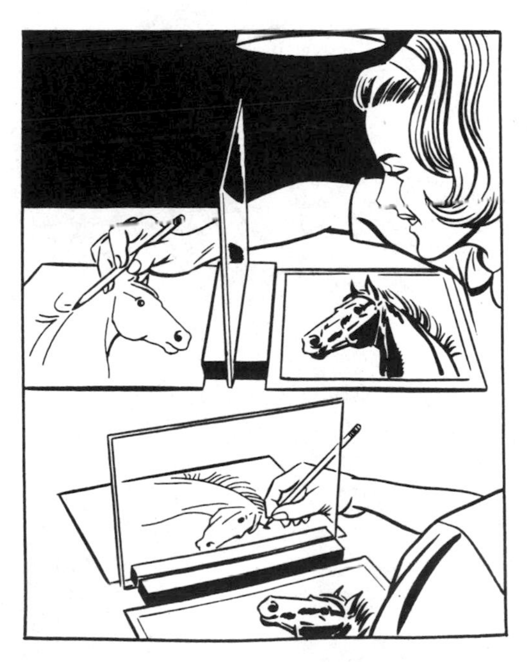

Do This: Place the picture to be copied in front of the glass and the paper behind it. An image of the picture will appear to be on the paper behind the glass and may be traced.

Here's Why: Both surfaces of the glass reflect some light. Some of the light reflected from the drawing is reflected from the glass surfaces to the eye. This makes the image appear to be on the paper behind the glass. The image is reversed, so the drawing traced from it will be in reverse.

NOTE: Various angles for lighting may be tried in order to get a clear image of the picture.

8

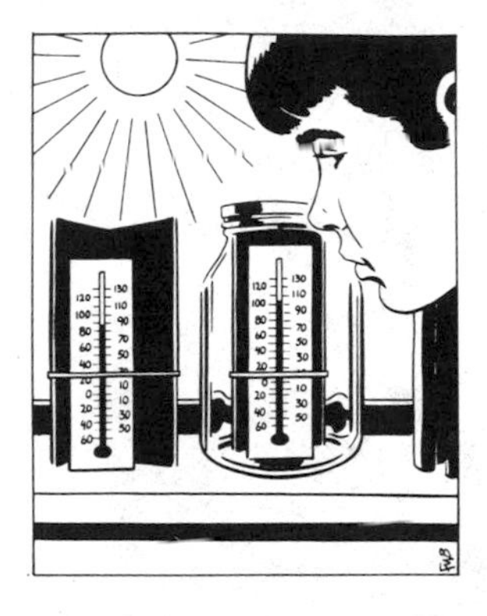

Heat

STEAM, VAPOR, GAS

Needed: A kettle of water heating on the stove.

Here's Why: Technically, we do not see steam, but only the tiny droplets of water formed as the invisible steam condenses into water. The droplets are so small that they rise with the rising air current. Bigger drops of water fall. Warm air rises over a hot pot because as it warms it expands and becomes lighter than cooler air.

Clouds may weigh many tons, yet they stay up because the small droplets of water of which they are made rise with rising air currents in the sky. If there are no air currents, however, droplets of water will fall slowly, regardless of their size.

NOTE. There is a difference between steam, vapor, and gas. When water boils, a white mist rises from it, but soon changes in air to an invisible vapor. The white vapor is not steam, and not a gas. It is composed of many molecules of water which aggregate together as the steam cools.

Steam coming from the spout of a kettle cannot be seen. It is invisible like most gases, but is not called a gas because its usual state is liquid. Oxygen, nitrogen, and carbon dioxide are examples of gases. They are in the gaseous form normally and can be made liquid only a very low temperatures.

HUMIDITY AND COMFORT

Needed: A small plastic bag, rubber bands, water.

Do This: Put a hand into the bag and put a rubber band around it at the wrist to make it airtight. Soon the hand begins to sweat and becomes wet. Wet the other hand with water at skin temperature. Now both hands are wet, but one is cool and the other is uncomfortable.

Here's Why: Perspiration normally cools the skin as it evaporates; so does the water on the hand. Here the air inside the bag soon takes up all the water it can hold so that no more can evaporate from the enclosed hand and none can escape from the bag. Then body heat causes the temperature to rise in the bag.

PUMP HEAT

Needed: A tire pump and two people.

Do This: Have someone operate the pump while another person holds one finger over the hose and the fingers of the other hand around the bottom of the pump. Note the heat and the cold.

Here's Why: Air when compressed is heated; the compression makes more molecules move in a given space. The warm air heats the metal pump cylinder. Friction of the pump plunger against the sides of the cylinder adds to the heat.

Expanding air is cooler; the air coming out of the hose expands and should feel cool to the finger. The sensation of heat is the bombardment of molecules. The faster they move the harder they strike against their container and the container gets hot (Charles's law).

HAMMER AND HEAT

Needed: A nail, hammer, solid piece of iron, such as an axe or a hatchet (a vise or anvil might be better).

Do This: Touch the nail to the cheek and it will feel cool. Place it on the iron, hammer it hard enough to flatten it somewhat, hold it against the cheek quickly and it will feel warm.

Here's Why: The moving hammer has kinetic energy (energy of motion). When it hits the nail, some of this energy is used as work to flatten the nail, some is converted to sound energy which is heard, and some is converted to heat energy as the flattening takes place.

The heat energy causes a temperature increase in the head of the hammer, in the solid piece of iron, and in the nail. The temperature increase is more noticeable where most of the work is done, in the flattening of the nail.

HEAT AND WEATHER

Needed: A bucket of water, bucket of soil, two thermometers, sunlight.

Do This: Insert the thermometers in the soil and water, and place the containers in the sun. Notice that the soil absorbs heat more rapidly than the water, as indicated by the more rapid rise of its temperature. The soil also loses heat more rapidly if the containers are put into a cold place.

COMMENT: This accounts for the usual changes of wind direction at the seashore. By day the warm shore warms the air above it. Then this warm air rises, bringing in a breeze from the sea. At night the shore loses heat. Therefore the water is warmer, the air above it rises, and the breeze reverses its direction, blowing from shore to sea.

FREEZE WITH FINGERS

Needed: Two ice cubes.

Do This: Press the ice cubes together, one flat surface tightly against the other. The cubes will freeze together.

Here's Why: The increase in pressure lowers the melting point and some of the ice melts at the place of contact. Then the water freezes again as the pressure is reduced.

COMMENT: Glaciers flow because of this principle. Where there is great pressure against the ice it melts a little, then freezes again after the pressure is lessened. This continued thawing and refreezing may let a glacier move a few inches a day.

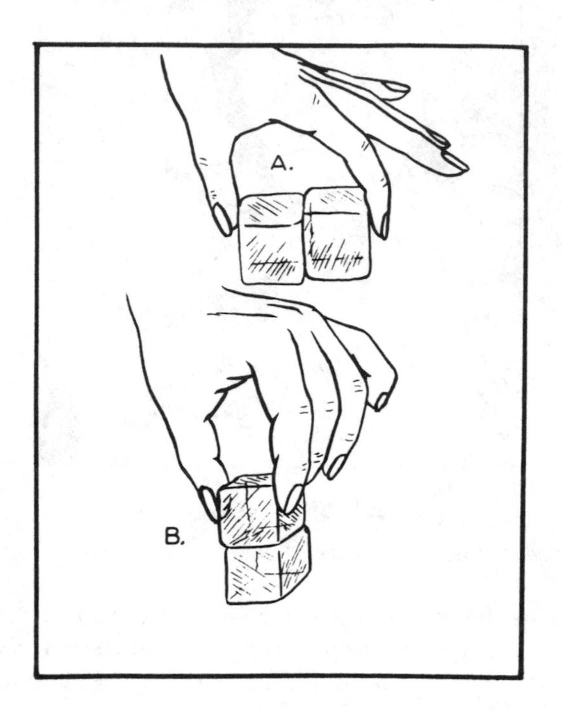

REGELATION

Needed: Ice cubes, a rack made of sticks and books to hold the cubes as shown, wires of different kinds, bricks, a pan to catch the water.

Do This: Tie the bricks to the wires and suspend them on the ice cubes as shown. The wires will cut through the ice cubes, yet the cubes will remain frozen together.

Here's Why: The pressure of the wires melts the ice because an increase in pressure lowers the melting point of ice. The water freezes again above the wires where the pressure is reduced. This is regelation. Note that smaller wires cut through more quickly because they exert greater pressure.

The flow of heat to produce regelation in this experiment is shown in the diagram of a cross section of a wire.

Regelation is important to all of us. It explains the flow of glaciers to the sea. Were it not for this property in ice, much more of the earth's water would concentrate at the poles, leaving the earth made up of ice and desert.

HEAT CONDUCTION

Needed: A candle, iron coat hanger wire, aluminum clothesline wire, six paper clips.

Do This: Twist the ends of the aluminum and iron wires together, lay them on a board, and drop candle wax on the clips to hold them to the wires. After the wax has cooled, hold the wire joint in the flame and watch as the clips fall off.

Here's Why: The clips on the aluminum wire drop off faster because aluminum carries or conducts heat better than iron. Try copper and other wires. The rate of conductivity of heat by different metals or other materials is called its specific conductivity and is rarely the same for different metals or alloys. It is interesting to note that in most cases the metal that conducts heat best also conducts electricity best.

BLANKET WARMTH

Needed: Two blankets and two thermometers.

Do This: See that the thermometers register the same temper-

ature. Wrap one in a blanket; the temperature does not change. Wrap a person *and* the thermometer in the other blanket; the temperature rises.

Here's Why: A blanket does not produce heat unless it is electric, therefore the thermometer wrapped in the blanket alone shows no increase. The blanket slows down the escape of heat from a warm body so that the thermometer wrapped in a blanket with a person shows an increase, the warmth coming from the person and not the blanket.

HEAT OR TEMPERATURE?

Needed: Two pints of tap water, each in a quart jar, one pint of hot water, one-quarter pint of hot water.

Do This: Have the pint and quarter-pint of water at the same temperature. This is easy if both are boiling. Pour the pint of hot water into one pint of tap water, and pour the quarter-pint of hot water into the other pint of tap water. Test the temperature with the hands. The temperature of the first mixture is much higher than the temperature of the second.

Here's Why: The heat of a body is the energy of motion of all its molecules added together. Its temperature is a measure of the average energy of motion of its molecules—the higher the temperature the faster they move around. Thus, the pint of hot water had more heat content than the quarter pint because it had four times as many moving molecules, even though its temperature was the same.

When hot water (fast molecules) and cold water (slow molecules) are mixed, the fast molecules bump into the slow ones, speeding them up, but slowing themselves down. The final temperature of the mixture depends on how many fast molecules are added as well as on how fast they are.

CONVECTION CURRENTS

Needed: A cooking vessel, water, and some rice.

Do This: Place the vessel of water off-center on the burner, and heat the water. Drop in a few grains of rice and they will rise on the side of the container over the heat and descend on the cooler side.

Here's Why: Water, when heated, expands and becomes less dense than the cooler water on the opposite side of the vessel. This causes the hot water to rise, while the heavier cooler water flows over to take its place.

When the water finally boils, rising steam bubbles add to the movement of the water. This cooling and heating process continues, and the motion of the water is a convection current. Arrows in the drawing show the direction of the movement.

THE STICKING ICE TRAY

Needed: A tray of ice just out of a cold refrigerator.

Do This: Note that the tray will stick to the fingers. An ice cube may stick to the fingers when it is first removed from the tray.

Here's Why: If the tray and cube are below the freezing point of water, the warmth of the hand will melt a thin layer of the ice or frost. Then, as the hand is cooled, the layer of water will freeze again.

It is possible that the hand or finger can freeze so tightly to the tray that a little skin is torn as it is pulled loose.

BOIL WATER WITH ICE

Needed: Saucepan, salt, water, pint fruit jar with tight lid, stove, ice cubes.

Do This: Fill the jar halfway with water. Place it in the pan and fill the pan with water almost as high as the water in the jar. Place the jar lid on loosely. Boil the water in the pan, adding salt until no more will dissolve.

Boil fifteen minutes, take the jar from the salt water, tighten the lid quickly and place an ice cube on the lid. Rub ice down the sides of the jar. The water in the jar will be seen to boil slowly.

Here's Why: Salt increases the boiling point of water in the pan, so that it can make the water in the jar boil. As it boils, the steam forces most of the air from the jar, taking its place. When the ice is applied, the steam is cooled. It condenses, reducing pressure on the surface of water in the jar. Reducing the pressure reduces the boiling point of water, and so the water in the jar boils.

MATCH VERSUS CHARCOAL

Needed: Wooden matches, piece of charcoal.

Do This: Strike the match, and blow out the flame. Kindle the piece of charcoal and try to blow it out. When breath is blown on a burning match, the flame is blown out. Blow on a charcoal fire and it burns hotter.

Here's Why: Cool air blown on the match surrounds the small area and cools it below the kindling temperature. Air blown against the hot charcoal meets a larger and more porous surface and does not have as much cooling effect. The oxygen in the blown air unites with the carbon of the charcoal, increasing the rate of burning and the heat.

Burning of the match is mainly a surface action, but burning charcoal involves much interior action.

ENERGY TO MELT ICE

Needed: Stove, pans, ice cubes, water, thermometers.

Do This: Put ice cubes in water and let the water temperature go down to near freezing. Use an equal amount of ice water in one pan and a mixture of ice cubes and ice water in the other. Put the pans on a medium hot stove, place the thermometers in the solutions and watch the temperatures change.

What Happens: Heat energy of the stove will raise the temperature of the ice water steadily. Heat energy supplied to the other pan is used mostly in melting the ice, and not until the ice is melted will the temperature rise noticeably higher.

The energy required to melt the ice is called the heat of fusion of ice.

THE DEW POINT

Needed: A bright, shiny tin can, water, small pieces of ice, thermometer.

Do This: Put water and the thermometer into the can. Then add ice slowly, stirring constantly. Watch carefully to see when the bright metal gets foggy and note the temperature at that instant. This temperature is the dew point temperature.

Here's Why: Dew point is the temperature at which water condenses out of the air and forms droplets. The dew point temperature is not a fixed temperature as is freezing point temperature for water but depends on the moisture content of the air. Put in salt and ice, continue to stir, and the dew on the can will freeze to form frost.

AN ICE PUZZLE

Needed: A transparent milk jug, crushed ice, water.

Do This: Fill the jug half full of ice, finish filling with water. Push the top pieces of ice down to water level and mark the jug at that level. Will the surface raise or fall as the ice melts?

Here's Why: Water contracts when cooled until the temperature reaches about 39 degrees Fahrenheit. Then it begins to expand. As the ice melts the water surface will go down. Then as the water rises to room temperature it will expand a little, and the water surface will rise. Marks on the jug with a grease pencil will show the different levels.

When the water and ice are first mixed, wrap a turkish towel around the jug for heat insulation. Then the ice can cool the water practically to the freezing point. If ice is allowed to melt in an open container of water, where it floats free of any downward pressure, the height of the water surface will remain the same when the temperature remains the same. See top illustration, page 142.

A HOT WATER MYSTERY

Needed: A jar or jug with tight-fitting lid or stopper, hot water.

Do This: Put a little hot water into the container, tighten the

lid and shake the container. Loosen the lid and a sizzling sound will be heard as pressure is released.

Here's Why: When the lid is tightened, the air in the jug is cooler than the water since there is a poor heat transfer from water to air. Shaking the jug cools the water a little and heats the air a lot.

Water has a very high specific heat as compared to air, which means that more heat is needed to raise its temperature a degree. So, as the air is heated the water is cooled only a little. The air expands greatly while the water contracts very slightly. This produces the pressure. Additional pressure may be caused by evaporation of some of the hot water, which becomes vapor in the jug. See bottom illustration, page 142.

WHICH COFFEE COOLS FASTER?

Needed: Two cups coffee, cream.

Do This: Put a definite amount of cream in the steaming coffee. Wait 15 minutes. Then put the same amount of cream in the other cup. Which will be cooler?

Here's Why: Sir Isaac Newton, some 300 years ago, discovered

that the cooling rate of a body is proportional to the temperature difference between the body and its surroundings. Therefore, cooling proceeds faster when the coffee is hotter. At the end of the experiment the cup receiving the cream last should be cooler than the other.

WAX "SNOWFLAKES"

Needed: A lighted candle and cool water.

Do This: Drop wax from the candle into the water. The wax will float on the surface, and its appearance will suggest snowflakes.

Here's Why: The hot wax is liquid, and when it comes into contact with the water it spatters and cools into irregularly shaped solid films.

The wax films suggest snowflakes in appearance only. A snowflake is a crystal and is much more beautiful than the blobs of wax formed in this way.

THE GREENHOUSE EFFECT

Needed: A large jar, two thermometers, two cardboards, bright sunlight.

Do This: Make easels out of the cards so they will hold the ther-

mometers upright. Place one in the jar in the sunlight, the other in similar sunlight,but not in a jar. Place them so the sun does not shine directly on the thermometers. The temperature in the jar goes higher than that shown by the thermometer outside the jar.

Here's Why: Short infrared or heat rays from the sun come through the glass and warm objects on which they fall. The warmed objects then give off infrared rays of a longer wave length. The shorter waves travel through the glass of the jar; the longer waves do not. So, their heat is trapped in the jar, while it is carried away from the thermometer outside the jar to other parts of the room. This is how the sun warms a greenhouse.

9

Biology

MYSTERY MOTION

Needed: A weight on a rope or string.
Do This: Hold the string with the weight suspended below it.

Think strongly that the weight will begin to swing back and forth, and it probably will. Think about it swinging in a circle and it probably can be made to do so without visible motion of the hand.

Here's Why: In a phenomenon called "body feedback" the muscles are directed by the brain to move exactly in the right way to start the object moving. Only slight movements at the right time make the weight swing in a wider and wider arc.

This movement can be directed consciously or sometimes subconsciously. If one consciously causes the weight to swing, note that it is not quite as easy to move the hand just right to bring the weight to a stop.

Our bodies have built-in positive feedback; we do not have built-in negative feedback.

NOTE: Do not expect as much action as that shown in the drawing. The drawing is somewhat exaggerated except for certain people.

WATER CIRCULATION THROUGH PLANTS

Needed: A stalk of celery, a glass of water colored with red ink or food coloring.

Do This: Place the celery stalk in the red water, and let it stand until the next day. Examine the celery.

What Happens: The red color will be seen in the leaves and stalk, showing the paths of the water through the stalk into the leaves.

Water rises continually from the roots of a plant to the leaves, where it evaporates into the air. The water brings nutrients up from the soil to nourish the plant.

THE GREAT REDI EXPERIMENT

Needed: Three wide mouth jars, three pieces of raw meat, paper, cheesecloth.

Do This: Place a piece of meat in each jar. Cover one with paper, one with two thicknesses of cheesecloth, and leave the other uncovered. Leave the jars out of doors in warm weather. The meat in all three jars will putrefy, but maggots will appear only in the uncovered jar.

COMMENTS: Aristotle, the Greek philosopher, taught that maggots appear spontaneously in putrefying flesh or filth. In 1668, Francisco Redi performed this experiment in a much more elaborate fashion, proving that life can come only from other life. In this case, maggots can come only from the hatching of fly eggs.

I found it necessary to protect the meat from night-prowling cats by enclosing the open jar in a wood frame covered with large-holed wire mesh.

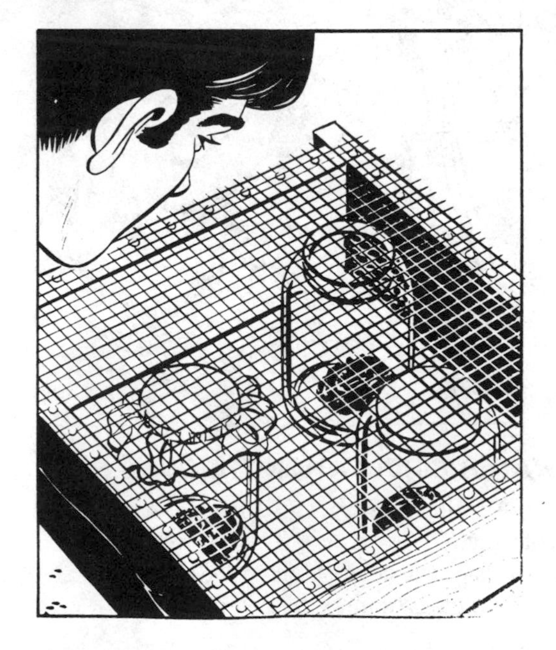

TRANSPIRATION

Needed: A large, thick plant leaf complete with all its stem, two glass jars, cardboard, modeling clay, water, sunlight.

Do This: Make a hole in the cardboard, place a stem through it and seal it with clay. Place the card over a jar of water so that the stem reaches into the water and place the second jar over the leaf. Put the jars in the light.

What Happens: Water traveling up through the stem is evaporated from the leaf and may be seen to collect in small drops inside the upper jar. Sealing the stem in the cardboard insures that no water will evaporate directly from the lower jar into the upper one.

The process by which leaves give off water, evaporating it into the air, is called transpiration.

ROOTS HOLD THE SOIL

Needed: A plant that has been growing for some time in a pot.

Do This: Hold a hand around the stem of the plant and dump it out of the pot. Notice that the soil will cling together in a mass the shape of the pot, and very little of it will fall off.

Here's Why: Roots extend themselves through the soil in which the plants grow, locking the soil particles so that they are not

washed away easily by the rain. Trees and plants hold our valuable soil. The soil washes away easily from our forests and gardens after plants are removed or killed.

A COCONUT CULTURE

Needed: A coconut.

Do This: Break the coconut. Expose the inside of it to the air

for an hour or more. Then close it again so it does not dry out. In a few days it will sour, and cultures of mold or bacteria in several colors will appear on the white meat.

Here's Why: The air is always full of bacteria and molds. Some of them settle on the moist coconut meat and start to grow there. The factors necessary for a culture of bacteria or molds are warm temperature, a food supply, protection from bright light which may kill them, moisture, and of course the exposure to air which provides the parent molds and bacteria.

SUNBURNED VEGETABLES

Needed: Growing potatoes, carrots, beets, or other underground vegetables.

Do This: Uncover some of the vegetables so that they are partly exposed to the sun but still mostly underground. The exposed parts turn green, or "sunburn."

Here's Why: Almost any plant exposed to light turns green. The green is chlorophyll, the catalyst in photosynthesis, the process by which plants manufacture starch and sugars from water and carbon dioxide. Light furnishes the energy for this process, and most plants will take advantage of light for this purpose by furnishing chlorophyll in the places exposed to the sun.

NOTE: The green parts of underground vegetables have an undesirable taste and contain small amounts of hydrogen cyanide, a poison.

OSMOSIS

Needed: Two slices of fresh apple, turnip, or carrot; some salt.

Do This: Place the slices on a plate and sprinkle a little salt on one. In a few minutes the salted piece will be covered with water and will be limber. The other, without salt, will remain stiff and may show signs of drying.

Here's Why: Osmosis here is the flowing of water through the membranes of the cells of the vegetables on which salt has been sprinkled. In normal osmosis, the greater flow is always in the direction of the denser liquid. Here the salt dissolving in the moisture of the vegetable increases its density, and the osmotic flow starts.

NOTE: In most cases vegetables should be cooked before they are salted. They are more tender and need less cooking time.

EXPERIMENT IN FOLKLORE

Needed: Mint leaves and ants.

Do This: Crush mint leaves in the fingers and draw a circle on the walk with them. Place an ant in the circle and it will refuse to cross the line.

COMMENT: I used horsemint for this experiment and it worked. I finally picked up the ant and set it out of the circle. A larger ant did cross the line after several hesitations.

According to folklore, mint growing around a house will repel ants, fleas, and other bugs. Rubbed on a dog, it will rid him of fleas. Mint is easy to grow in most localities, and a thorough experiment with it would make a science fair project.

SPLITTING STALKS

Needed: Rhubarb.

Do This: Put the rhubarb stalks in water. Soon the ends will begin to split and curl.

Here's Why: Rhubarb stalks consist largely of elongated vertical cells of several different kinds of groups known as fibrovascular bundles. These are associated with other cells of thinner walls and less vertical elongation.

When the stalks are placed in water, some water is absorbed by osmosis and the fibrovascular bundles split vertically along planes of least strength. The curling results because some of the cells absorb more water than others, and so expand more.

COLOR FATIGUE

Needed: Bright red and bright yellow paper, a white sheet of paper, good daylight.

Do This: Look steadily at the red paper for two minutes, never taking the eyes away. Then look quickly at the white paper. The white paper will look bluish-green. Stare at the yellow paper the same way. Then look at the white, and a blue color will be seen. *Here's Why:* One theory of color vision assumes that in the eye there are three types of retinal cones. One type is very sensitive to red, one to green, one to blue. By staring at the red paper in bright light the red-sensitive cones become fatigued or tired and do not function properly for some little time. The white paper reflects all colors from its surface, but appears bluish-green because the red-sensitive cones are tired.

HOW PLANTS BREATHE

Needed: A house plant and a jar of vaseline.

Do This: Coat the top sides of two leaves and the undersides of two other leaves with vaseline. Leave them on the plant for one or two days. Then compare. The leaves coated on the lower sides will be dying.

Here's Why: The openings, called stomata, through which the plants take in air containing carbon dioxide, are on the undersides of the leaves. The vaseline closes the stomata so that air cannot enter.

Light used by the plant to make its food shines mostly on the upper sides of the leaves. The vaseline coating does not cut off the light. Another ingredient, water, comes from the roots through the stems; it too, is not affected by the vaseline.

HOT AND COLD

Needed: A bowl of very warm water, a bowl of ice water and a bowl of tap water.

Do This: Put one hand in the warm water and one in the ice water. Then remove both hands and plunge them into the tap water. The tap water will feel cold to the hand that was in the hot water and warm to the hand that was in the ice water.

Here's Why: The hand moved from the warm water to the tap water quickly loses heat. The hand moved from the ice water to the tap water quickly gains heat. The sensation of hot and cold depends on the direction and rate of heat flow.

A SPOREPRINT

Needed: A large mushroom, cardboard or paper.

Do This: Place the mushroom on the paper and leave it overnight.

What Happens: Spores will have fallen out of the mushroom and formed into a "print" on the paper, outlining the structure of the underside of the mushroom.

A spore is a tiny one-celled organism that can grow into a new plant, much like other plant seeds do. There are millions of spores in an ordinary mushroom.

MOTHER OF VINEGAR

Needed: Grape juice or some other non-citrus juice, a warm place, plenty of time.

Do This: Let the juice ferment. After a few months a slimy mass may be seen in the liquid, which is now about four percent acid vinegar.

Here's Why: Yeast plants from the air multiply in the juice, changing sugar to grain alcohol. Microscopic rods of acetic bacteria in the presence of oxygen from the air convert the alcohol into vinegar.

The slimy mass composed of rods of the bacteria, potassium carbonate, and other mineral salts from the juice, is called "mother of vinegar." It is of no use, but old-timers, in making vinegar from hard cider, always put a little "mother" with the cider into a wooden cask and let it set six months to a year.

BAD AIM

Needed: A card with a hole in the center, a pencil.

Do This: Hold the card and pencil as far apart as possible, close

one eye, and try to put the pencil into the hole quickly. Most people will miss the hole almost every time.

Here's Why: We are accustomed to seeing objects with two eyes, which gives depth to our vision. Our eyes measure distances by triangulation, as engineers measure distances very accurately by sighting from two different points through their transits. When we are deprived of this faculty our vision is not as accurate.

When observing a nearby object, the eyes are turned toward each other, but when we look at a distant object, they aren't as cross-eyed. Our brains performs quick mathematical tricks, measuring the turning of our eyes, and tells us the distance just as engineers can figure it from their transit measurements.

THE BERLESE FUNNEL

Needed: Cardboard, wood, wire mesh, string, a jar of rubbing alcohol.

Do This: Build the funnel as shown. Place freshly dug moist soil and humus on the mesh. A collection of small night-crawling insects will fall into the alcohol.

Here's Why: Antonio Berlese, a twentieth-century entomologist, invented this funnel to collect insects for study. As the soil dries out, the insects travel toward the darkness of the funnel and fall into the alcohol.

THE DANDELION

Needed: A dandelion flower and a magnifying glass.

Do This: Examine the flower to find its parts. You'll find that the flower is a collection of many tiny flowers, each complete with pistils and stamens and each capable of developing a seed that can grow into another plant.

To find the number of flowers in the dandelion blossom it is easier to count the seeds after they have developed on their silky wings.

COMMENT: Such a flower is called a compound flower. There are hundreds of others, but perhaps none as bothersome on a lawn as the dandelion. Many beautiful commercial flowers are members of the same family, the composites.

TASTE BUDS' LOCATIONS

Needed: Toothpicks, sugar water, vinegar, fruit juice.

Do This: Dip a toothpick into the sugar water and touch it to different parts of the tongue. The sweetness can be tasted at definite locations on the tongue, mainly at the tip.

Dry the tongue with tissue or a cloth, place some dry sugar crystals on the tip, and the sweetness is not noticed. A drop of wa-

ter placed on the crystals enables us to taste them.

Here's Why: Different taste buds respond to different tastes, and these are located in different parts of the tongue. The back of the tongue, for instance, is most sensitive to bitter tastes.

Some substances are not tasted unless their odors reach the nose. Our sense of smell gives us much of our ability to taste.

10

Electricity and Magnetism

ELECTRICAL COLOR CHANGE

Needed: Blue ink, salt, water, nine-volt transistor battery, wire, very small glass container, larger container for mixing.

Do This: Put water into the large glass, and add as much salt as will dissolve. Pour some of the salt water into the small dish. Add a drop or two of ink to make a blue color. Connect wires to the battery and place the bare ends of the wires in the solution. Bubbles will form around the wires and the color of the solution will begin to change.

Here's Why: Tiny amounts of sodium hydroxide are formed at the negative wire. Chloride ions are formed at the positive wire, and these react with water to form HC1O, which will not hold together. It breaks up into HC1, hydrogen chloride, and a very active form of oxygen we call "nascent." The nascent oxygen is searching for another element or substance to combine with, and the ink colors are its victims. It destroys them.

At very small currents the chlorine is consumed by reaction with the copper wire or ink. With higher current both chlorine and oxygen come from the positive electrode. The sodium hydroxide also can attack and destroy ink color. So, in a simple way we can produce complex chemical reactions.

NOTE: I used Sanford's Penit blue-black ink. Some other inks did not work.

INK
BATTERY
FWB

FWB

MYSTERY LIGHT

Needed: A roll of electrician's friction tape, a dark room.

Do This: Stay in the dark until your eyes are accustomed to it. Then look closely as the tape is pulled off the roll quickly. A light is seen where the tape pulls off.

Here's Why: The light is electric, although the reason why is not fully understood. As the tape is stripped off and the adhesive bond broken, there is a charge separation. The adhesive side takes on a charge of one sign, and the other side takes on the opposite charge. The charges leak back through the air to neutralize each other. As the charges pass through the air they collide with air molecules, exciting them to give off light. See bottom illustration, page 163.

READING THE METER

Needed: A water, electric, or gas meter.

Do This: Make a drawing of the dials and the positions of the hands. Make another drawing in one month. The difference in the readings indicates the kilowatt-hours of electricity, cubic feet of gas, or gallons of water used during the month.

NOTE: Be very careful in reading the dials. When a hand points almost directly toward a number, observe the dial at the right of

it. If the hand at the right has passed 0, the previous dial reading is the number to which the hand points. If its hand has not passed the 0, the previous dial reading is one less than the number to which the hand points.

In the drawing, the correct reading is 3,632.

AN EDDY CURRENT BRAKE

Needed: Wood, tools, a nail, strong magnet, copper.

Do This: Build a device in which a wooden strip can swing freely on the nail. Attach copper strips (I used three ounces of copper) to the bottom. Start it swinging and note how long it will swing. It will stop more quickly if the magnet is held close to it.

Here's Why: Electrical currents are induced into the copper as it swings through the magnetic field of the magnet. These "eddy" currents in turn produce magnetic fields which act in opposition to those of the magnet which tends to act as a brake.

Iron parts of electric generators, motors, and transformers are made in thin sheets to reduce eddy currents which are undesirable. They have a good use, however. In electric kilowatt-hour meters they act to stabilize the moving parts.

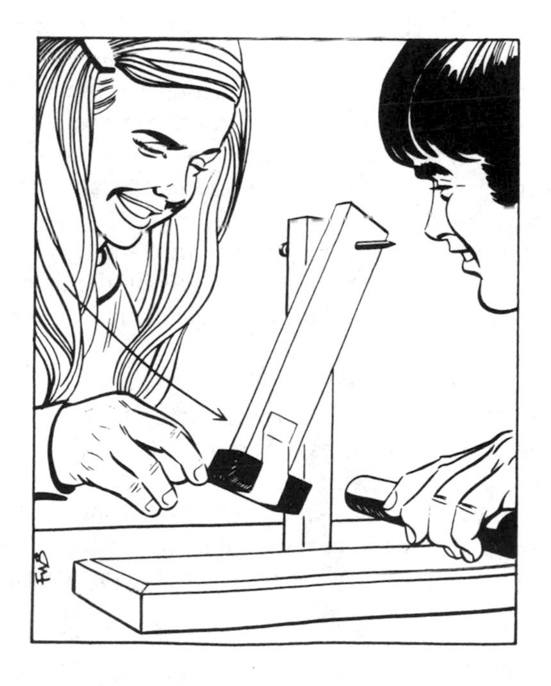

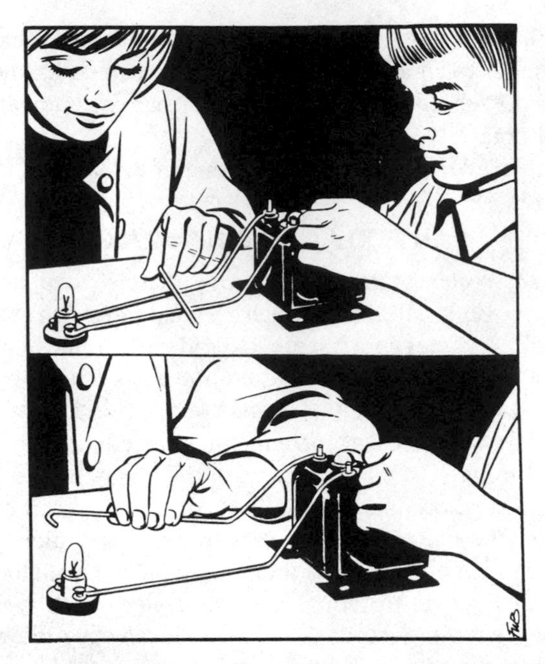

OPEN CIRCUIT, SHORT CIRCUIT

Needed: Toy transformer or battery, wire, a light or small motor.

Do This: Connect the power source to the light or motor and the power flows smoothly to the right place. Drop another wire across the two used to make the connection and the light goes out or the motor stops. This is a short circuit; the current takes an easier path and does not reach the light. The word "short" does not mean shorter length, but an easier and undesirable path for the current.

Restore the power by removing the wire and this time disconnect or cut one of the connection wires. Again the light goes out; this is an open circuit. The path of the current is incomplete.

AC AND DC

Needed: Aluminum pie plate, cotton cloth, starch or flour, potassium iodide, water, two wires, a battery, a toy train or bell transformer.

Do This: Make a thin paste of the starch and the potassium iodide (a small amount from the drug store), soak the cloth in it, and place the wet cloth over the bottom of the inverted aluminum pie plate.

Connect one wire from the negative pole of the battery to the aluminum. Connect the positive terminal of the battery to the other wire, hold it in the hand, and rub it across the cloth. A blue streak will be seen on the cloth.

Now connect the wires to the transformer instead of the battery, rub one wire across the cloth, and a dashed line will be left on the cloth.

Here's Why: Electric current neutralizes the negative iodide ions to liberate free iodine molecules. These combine with starch to form blue starch iodide. Iodine is liberated only when the wire touching it is positive. Electricity from the transformer is ac which reverses direction in the wires 120 times per second. Thus, the wire rubbed across the cloth is alternately positive and negative, and leaves its mark only when it is positive. The battery gives dc which does not reverse. (Ac means alternating current; dc is direct current. The normal frequency of ac in the United States is 60 cycles.)

A SIMPLE ELECTROMAGNET

Needed: A battery or toy transformer, insulated wire, iron straight pins or paper clips.

Do This: Connect the wire to the terminals of the power source, turn on the current, and see if the magnetic field around the wire will pick up any pins. The wire may get hot, but probably no pins will stick to it.

Make a few turns of the wire around a pencil, run the current through it, and the field will be stronger. It may pick up a pin. Next, wrap a few turns around a small iron bolt and the resulting magnet will pick up the pins and clips.

Here's Why: An electric current traveling through a wire will produce a weak magnetic field which may be detected with a compass. When the wire is in the form of a coil, the field will be stronger. It will be much stronger if a piece of iron is inside the coil.

A MAGNETIC SWING

Needed: A horseshoe magnet, thin razor blade or small nail.

Do This: Place the metal between the poles of the magnet so that it touches one and is close to the other. Give it a push and it will swing back and forth.

Here's Why: The metal object becomes an extension of the upper magnetic pole to which it is attached, and its lower end is attracted to the bottom pole of the horseshoe magnet.

I used a 20-penny nail sawed off so that the point of the nail could be used as the fulcrum. Friction was reduced to a minimum and so the nail, when pushed aside, would swing back and forth as if controlled by a spring.

NOTE: If the upper pole of the magnet is north, the upper part of the nail becomes south and the lower part of the nail becomes north. It is attracted to the lower pole of the horseshoe magnet, which is south.

THREE-DIMENSIONAL LINES OF FORCE IN A MAGNETIC FIELD

Needed: A strong magnet, piece of window glass, small nails.

Do This: Place the glass over the magnet and let the nails drop to it one at a time. The nails will make a pattern indicating the direction of the lines of force in the magnetic field.

When the glass is inverted, most of the nails will cling to the bottom of it, as shown.

Here's Why: In the presence of the magnet the iron nails become temporary magnets. The nails line up in the direction of the magnetic field of the magnet, just as a magnetic compass needle lines up in the direction of the earth's magnetic field. If a nail is not lined up, the magnetic lines will pull it into line.

If the assembly is inverted, the nails will still indicate the pathways of the lines of force.

THE PINCH EFFECT

Needed: A number of parallel wires, a switch, a source of current such as a battery or transformer, some combs and wood.

Do This: Turn on the switch and the wires attract each other near their centers.

An easy way to string the wires and change them to different patterns is by using combs stuck into notches in a piece of wood. Set up as shown.

Here's Why: A wire carrying a current produces a magnetic field around it. If the current in parallel wires is traveling in the same direction, the magnetic fields produced attract each other, pulling the wires toward each other.

NOTE: In experiments aimed at creating a controlled fusion reaction to produce atomic energy, a plasma, an extremely hot gas

at millions of degrees temperature, is used instead of wires. Attempts have been made to use the pinch effect to keep the hot gas in a "magnetic bottle" so it will not vaporize the wall of the container. So far the experiments have not been successful.

INDUCED CURRENT

Needed: Two copper wire rings suspended by strings, a strong magnet.

Do This: Solder the ends of one ring together; leave a gap between the ends of the other. As the magnet is pushed into the soldered ring, the ring moves in the direction of the push; as it is withdrawn the ring swings backward toward the magnet. Motion of the magnet has no effect on the unsoldered ring.

Here's Why: As lines of force from the magnet pass through the soldered ring, an electric current is produced—induced—in it. It becomes a magnet and reacts with the moving magnet according to the rule: like poles repel and unlike poles attract. No current can be induced in the ring that does not have a continuous path for the current (Lenz's law).

NOTE: My rings were made of 12-gauge copper wire. Iron wire will not work because it is attracted by the permanent magnet.

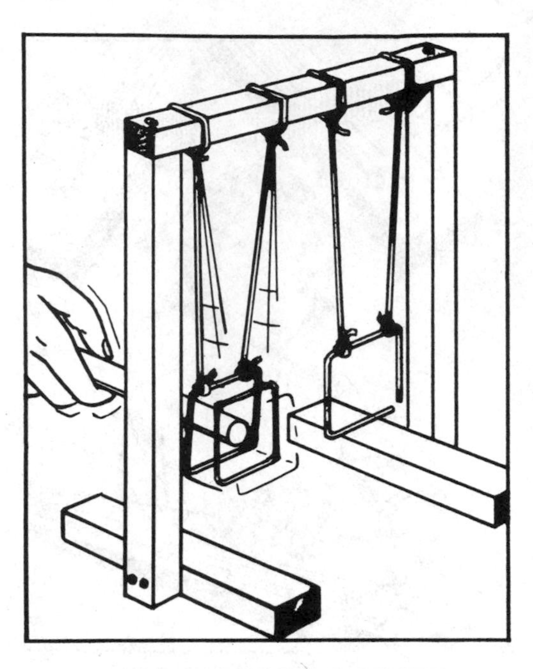

DISTORT THE TV PICTURE

Needed: A strong magnet, an old television set in operation.

Do This: Hold the end of the magnet near the picture and the

picture is distorted. Move the magnet about and note the "flowing" of the distortion.

Here's Why: The picture on the screen is made up of electrons which are shot from the back end of the picture tube to the front. Electrons are diverted from their straight path by a magnetic field.

Magnetic field deflection of charged particles is used in many important electronic devices, such as the mass spectrograph, the cyclotron, etc.

NOTE: Do not try this with a color television set, which is more easily put out of adjustment than a black-and-white TV.

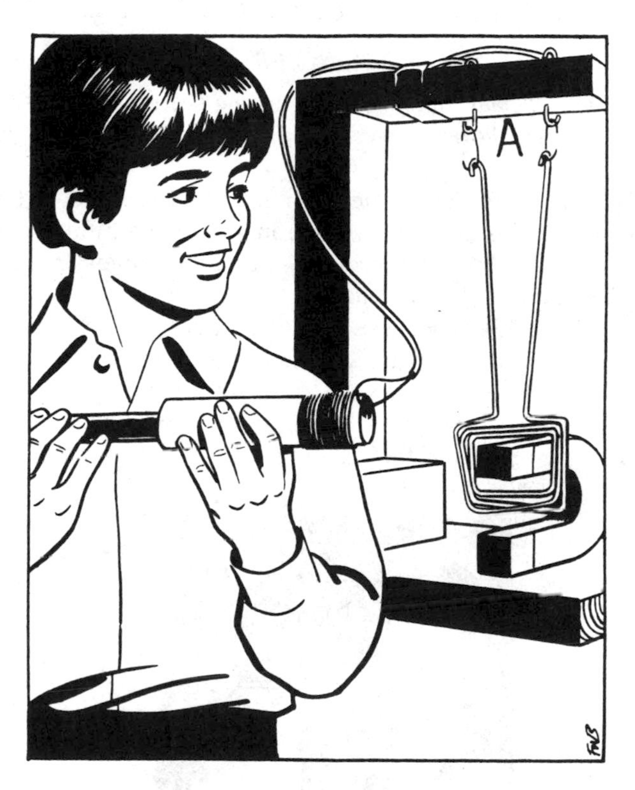

A MYSTERY PENDULUM

Needed: A *strong* magnet, magnet wire, a cardboard tube, lamp cord, wood and tools, soldering equipment, an ordinary magnet.

Do This: Construct the pendulum as shown. The bottom of the pendulum is made of three turns of magnet wire and is suspended from the lamp cord by single strands of wire from the cord to make for easy swinging. Wrap 50 turns of magnet wire around the tube.

When the strong magnet is moved in and out of the tube, the pendulum will begin to swing. If the magnet motion is timed rightly, the pendulum will be made to swing high.

Here's Why: Motion of the magnet in the coil produces electric currents in the coil. The current goes in one direction as the magnet is inserted and in the other direction as the magnet is withdrawn. The currents transmitted to the pendulum through the lamp cord make a magnet out of the pendulum, and this is alternately attracted and repelled by the U-magnet.

NOTE: I found that all connections, including those at A, had to be soldered in order for good contact to be made.

THE JIGGLING FILAMENT

Needed: A clear lamp bulb with a long section of straight filament, a magnet. To avoid eye string *very* dark goggles or sunglasses should be worn.

Do This: Hold the magnet near a straight section of glowing filament. The filament will vibrate in a very beautiful pattern.

Here's Why: Any wire, including the lamp filament, has a magnetic field around it when current flows through it. Since the ac reverses itself 120 times a second, the magnetic field around the filament also reverses itself 120 times a second.

The alternate attracting and repelling of the magnetic lines of the magnet and the magnetic field of the current-carrying filament causes the vibration of the filament.

INDUCTIVE MAGNETISM

Needed: A permanent steel magnet, a nail, some soft iron paper clips, screws, or tacks.

Do This: Hold the nail close to the magnet without touching it. The other end of the nail will pick up the clips but will drop them if the magnet is moved away from the nail.

Here's Why: The nail becomes a magnet by induction as the lines of force from the magnet pass through it. Since the nail is not hard steel, it loses most of its magnetism as the magnet is withdrawn.

MAKE A MAGNETIZER

Needed: Wood, a cardboard tube from bathroom tissue, tape, 70 feet of number 14 or 12 gauge house wire, screwdriver.

Do This: Put wooden pieces into the cardboard to prevent it from crushing. Thread tape through and put end pieces over the wood and against the tube to hold the wire turns. Wind the wire neatly and bring the tape over the coil. Tie it to hold the turns in place.

Disconnect a storage battery lead in an automobile and connect your coil in series with the lead and the battery terminal. Insert a screwdriver into the coil, have someone turn on the starter switch, and in ten seconds you have a magnetized screwdriver.

Here's Why: The heavy current from the battery coursing through the turns of wire produces magnetic lines of force. These extend through the steel which has been inserted inside the coil, rearranging the molecules in the metal so that it becomes a magnet.

NOTE: The coil used for the drawings was made to restore magnets that had lost their magnetism through long use in a school. It worked satisfactorily. Adult supervision is necessary for this experiment.

THE DIPPING WIRE

Needed: Fifteen feet of number 20 bell-wire, a metal lid, water, a strong electric current from a storage battery or a large train transformer.

Do This: Make a device such as is shown in the lower drawing and suspend it from the coiled wire. The point of the loose wire in the device should barely touch the lid. Connect one side of the power source to the lid and the other to the coiled wire. Place some water in the lid. The device suspended from the coil will move up and down.

Here's Why: As current flows through the coiled wire, the magnetism created moves the turns of the coil closer together raising the point from the lid, thus breaking the circuit. When the circuit is broken the coil releases again, and the point comes back down to make contact.

The water prevents the lid from burning. In a laboratory this is avoided by having the point of the wire dip into mercury. An iron bolt mounted inside the coil will enable this device to work better on less current.

THE MOTOR EFFECT

Needed: A wooden frame, clean copper or aluminum wire, small connecting wire, battery, magnet.

Do This: Set up the apparatus as shown, allowing the bottom ring to swing freely. When current is passed through the ring it swings outward or inward, according to the direction of the current.

Here's Why: Danish physicist H. C. Oersted discovered that a wire carrying a current produces a magnetic field around it. M.

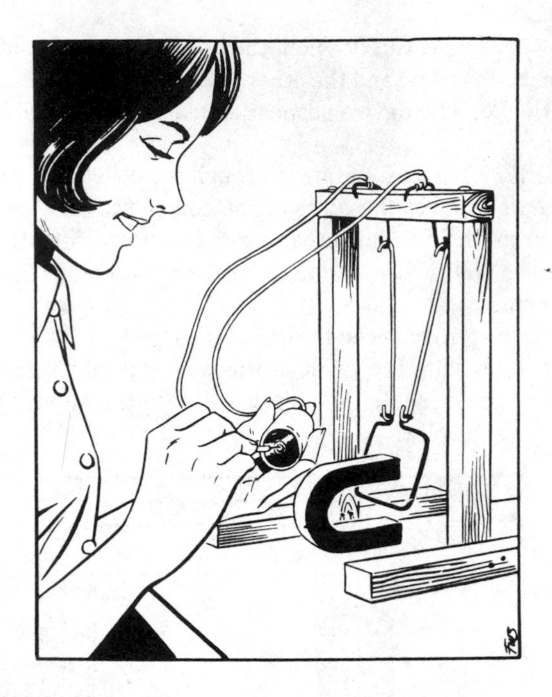

Faraday used this discovery to produce small movement with the current, as shown here, and later Joseph Henry produced the first appreciable movement from electricity—the first electric motor.

When current is passed through the ring it becomes an electromagnet and repels or attracts the permanent magnet according to the law: like poles repel and unlike poles attract. A strong magnet or a stronger current produces a stronger motion.

A RHEOSTAT

Needed: A flashlight bulb and socket, window shade spring, dry cells or transformer.

Do This: Make the connections as shown. The bulb should glow brightly when its wire is touched on the spring near the end where the other wire is connected but should grow dimmer as the wire is made to slide long the spring toward the opposite end.

Here's Why: The steel wire of the spring is not a very good conductor of electricity, and the more wire through which the current must pass, the less electricity can travel through it. This is a rheostat—a device to vary the amount of a current passing through it to complete a circuit. See top illustration, page 179.

SLOW THE FAN DOWN

Needed: Two sockets and a plug, lamp cord, electric fan, lamp.

Do This: Connect the sockets and plug as shown. When a lamp is put into socket A, and the fan is plugged into socket B, the lamp should burn dimly and the fan should run more slowly and quietly than usual.

Here's Why: Suppose the fan draws 40 watts of power. If a 40-watt lamp is put into the socket it is in series with the fan, and both lamp and fan will be getting only half the voltage of the line. Therefore, the fan runs slowly and the lamp burns dimly.

If the fan still runs too fast, use a smaller lamp. A larger lamp will let the fan run faster and bulb burn dimmer.

NOTE: Do not try this method to slow down other types of motors. Damage to the motor could result. See bottom illustration.

Index

F

G

H